WHY WE AGE

WHY WE AGE

What Science Is Discovering about the
Body's Journey Through Life

Steven N. Austad

John Wiley & Sons, Inc.

New York • Chichester • Weinheim • Brisbane • Singapore • Toronto

This book is printed on acid-free paper. ∞

Copyright © 1997 by Steven N. Austad. All rights reserved

Published by John Wiley & Sons, Inc.
Published simultaneously in Canada

This publication is designed to provide accurate and authoritative information in regard to the subject matter covered. It is sold with the understanding that the publisher is not engaged in rendering professional services. If professional advice or other expert assistance is required, the services of a competent professional person should be sought.

Library of Congress Cataloging-in-Publication Data:

Austad, Steven N.
    Why we age : what science is discovering about the body's journey
through life / Steven N. Austad.
       p.   cm.
    Includes index.
    ISBN 0-471-14803-2 (cloth : alk. paper) / ISBN 0-471-29646-5 (paper :
alk. paper)
      1. Aging—Physiological aspects.   2. Age factors in disease.
I. Title.
QP86.A97   1997
612.6'7—dc21   97-5542

Printed in the United States of America

10 9 8 7 6 5 4 3 2 1

*For Vince and Bettie, who never gave up on their
sometimes wayward son*

Contents

Preface

"You're writing a book on aging? How depressing. Well, hurry and come up with some answers so I know what to do before it's too late."

I get this response pretty often when people find out what I've been up to for the past couple of years. They must think me a sort of morbid voyeur, like someone who attends the funerals of strangers. But oddly enough, the topic is anything but depressing, as I hope I will show. Aging represents a biological paradox that few people appreciate, and for such a nearly universal process, it is nearly limitless in its variety. A mayfly lives a day, a fly a week, a dog a decade, a human a century, a tree a millennium or two. Salmon live a few years, then spawn and die within a few days. For turtles, to get older may be to get better. Is there a pattern here? What is it and why? Is the pattern fixed or can we alter it? Are there 150- or 160-year-old people in the world? Do men age more quickly than women? Do dolphins or gorillas get arthritis or Alzheimer's disease, or are we the only lucky ones?

In retrospect, I think I was attracted to aging research by its scope and the conceptual contributions that a comparative zoologist could still make. The questions seemed endless, but the answers achievable. If nothing else, I think I've convinced many gerontologists that animals besides rats and mice have stories to tell us about aging. And besides, everyone cares. In his book *Advice to a Young Scientist*, Peter Medawar suggests, "Study what interests people." How could you not be interested in what your mirror tells you daily?

Besides, since the Age of Discovery ended there has probably been no more exciting time to be a biologist. Palpable progress is being made on many fronts of basic research. The brain is finally yielding up its secrets. We are suddenly godlike in our ability to manipulate organisms' genes. An elm gene can be put into corn. We have bacteria producing human hormones. Soon we will have designer plants, manufacturing their own fertilizer and pesticides. Computers now use evolutionary principles to design new therapeutic drugs.

And yet, the biggest nut to crack of all of them is aging. Nothing would alter human life more dramatically (whether for better or worse isn't exactly clear to me) than our learning to delay aging—if we had an active working life of a century, if athletes remained in their prime for 50 years, if we could live with six or eight generations of our family, if we might live to suffer the long-term consequences of short-sighted political foolishness.

A few years ago, a prominent gerontologist lamented that aging research never seemed to make progress. There were too many ideas chasing too few facts, just as there had always been. The problem might well be too complex for us ever to understand. We might as well join the parade of cranks and charlatans peddling their longevity creams by the bucket.

But the gloom is gone. There is a subdued but palpable excitement in the air now when gerontologists meet. We are closing in on the fundamental processes of aging, learning what drives them. Next, we will begin to tinker.

This is not to say that hokum and charlatanry, or just simple wishful thinking, have disappeared. I admit to wincing a bit every time I read in the popular press that one of my colleagues has opined that, having discovered a gene that doubles the life span of a single-celled fungus, we are closing in on the 400-year human life span. Few of us believe this. One of the aims of this book is to separate substance and reasonable speculation from myth and wishful thinking. No matter what you read in the newspapers, the 200- or 400-year life span is not just around the corner, although some significant progress in slowing aging may be.

In writing this book, I hope to convey some of the colorful history of the field, the parade of ideas, the outsized personalities, and the key experiments and experimenters, as well as our current excitement and the reasons for it. To the extent that I'm successful, I would hope that both the lay and professional reader might learn to think about aging without tears or terror, to think about it as an intriguing puzzle rather than a gloomy inevitability. Nothing clarifies one's thoughts as writing does, and the process of crystallizing my thoughts on paper has been a revelation. E. M. Forster probably put it best: "How do I know what I think until I see what I say?" Now I know.

Parts of this book concern areas distant from my area of expertise. I've prodded and pestered my colleagues, the experts who would tolerate it, to assist me in these areas. For reading parts of the manuscript or answering off-hand questions of no obvious relevance at the time, I'd like to thank Gary Dodson, Donna Holmes, Tom Johnson, Veronika Kiklevich, Ed Masoro, Roger McCarter, Gerry McClearn, Jim Nelson, Jay Olshansky, Olivia Pereira-Smith, Michael Rose, Jim Smith, Dick Sprott, Huber Warner, and Peter Waser. For trying to keep me from sounding too dryly professorial, I'd like to thank my editor Emily Loose. My agent, Rick Balkin, helped pull all aspects of the book together. Thanks to Bill Day of John Hopkins Alan Mason Chesney Medical Archives for guiding me through Raymond Pearl's papers. And most emphatically, for enduring the whole manuscript and providing their massive expertise, special thanks are due to Caleb Finch, George M. Martin, and Richard Miller. Their comments and criticisms have improved the text immeasurably and saved me from more than a couple of bloopers. Any remaining mistakes are, of course, entirely due to my own efforts.

For supporting my research on aging over the years, I'd like to thank the U.S. National Institute on Aging, the U.S. Department of Energy, Harvard's Henry Rosovsky Fund for Research, and Idaho's State Board of Education.

Without the help of Mike Smith and Lehr Brisbin at the Savanna River Ecology Laboratory and Jim Alberts and Charles Durant at the University of Georgia's Marine Research Institute on Sapelo Island, my research on aging may have never got off the ground.

For commissioning the magazine article that inspired me to finally begin this book, I thank Rebecca Finnell of *Natural History*.

For putting up with my moods and muttering and absent-mindedness while I labored at writing, I'm grateful to my wife, Veronika, and my children Molly and Marika. I hope they all enjoy getting reacquainted with me now that I've finally stopped camping out in the study.

1

The Paradox of Aging

In Westminster Abbey, where England buries with solemn honor its greatest poets, painters, scientists, and statesmen, lie the bones of one rather ordinary man—Thomas Parr. A farmer's servant from Shropshire, Thomas Parr's only claim to distinction is that he managed to convince a gullible seventeenth-century public that he had been alive for more than 150 years.

As it turned out, the glory gained from such a claim was a mixed blessing. Although Parr did manage to have himself interred among the anointed paragons of English history, had he not made his claim to great longevity, he probably would not have been dragged in from the provinces to be exhibited before the king like an exotic plant from the New World. And if he hadn't been so exhibited, he might not have caught the illness that, a short time later, finished his mendacious history at the declared age of 152, sending him straightaway to rest at Westminster.

The story of "Old Parr" is instructive in several senses. Most fundamentally, it shows how desperately gullible we humans are about things we fervently wish to believe—such as the possibility of extremely long life. Thomas Parr was without question a charlatan, something that should have been obvious to minimally skeptical minds even then. How do we know? For one thing, the only evidence of his age was his own word about his birth date, and the fact that he appeared to be very old. Also, as a final posthumous honor, Parr was autopsied by William Harvey, the most famous physician of his day. In the autopsy report, Harvey noted that Parr's internal organs showed little sign of extreme age. That Parr's own account was accepted at the time is even more remarkable when one reads in the same autopsy report that

> his memory . . . was greatly impaired, so that he scarcely recollected anything of what had happened to him when he was a young man, nothing of public incidents, or of the kings or nobles who had made a figure, or of the wars or troubles of his earlier life, or of the manners of society, or the prices of things.[1]

So although Parr invented an acceptably intriguing biography for himself—bachelorhood until the age of 80, a second marriage at 120, and a child soon after, plus noble hard work in the fields until 130—humans, as we shall see later, do not live to be 150 years old or even 140 or 130, hard work or clean living notwithstanding. The Parr who is buried in Westminster Abbey was perhaps the son or even the grandson of the Thomas Parr who claimed to have been born in 1483. He was undoubtedly an elderly man when he died, but elderly then, as now, was 70 or 80 years of age, not 140 or 150.

We might like to think that we are not as gullible as the unsophisticated gentry of the seventeenth century, but all evidence points to the contrary. *Life* magazine, for instance, published an article in 1966 on an area in the Georgian Caucasus where the people commonly claimed to live for more than 100 years.[2] This story focused on Shirali Muslimov, the oldest man in his village, who claimed to be 161 years old at the time. He had married his current wife when he was supposedly 110, and still seemed hearty and energetic, although there was no mention of any children on the way. When he died seven years after the *Life* article appeared, the news was reported around the world. Similar stories have come out of Ecuador and northern Pakistan, and for a time in the mid-twentieth century, claiming extreme longevity became something of a cottage industry among remote villagers around the world when they were discovered by credulous anthropologists or journalists.

American newspapers in October 1979 were full of the news that a man named Charlie Smith, who claimed a boyhood memory of being brought to the United States as a slave in 1854, had died at the age of 137. Smith had a brief run in the *Guinness Book of World Records* as the longest-lived human in recorded history, and just before he died, he was the subject of a television show recounting his exceptionally long life. But Smith's claim was disproved shortly before his death when a marriage certificate surfaced that he filled out when he was 35 years old in 1910. In truth, Smith died at the respectable, though not unheard of, age of 104. Thus his claim, like others when subjected to sufficient scrutiny, was false.

In addition to illustrating the depth of human gullibility, the story of Old Parr illuminates what might be called the paradox of aging. It's apparent to anyone who cares to notice that the elderly in the seventeenth century lived, as now, in a state of relatively fragile health. And generally speaking, the more elderly they grew, the more fragile they became. As Peter Medawar, the Nobel Prize-winning immunologist, once put it, "What lays a young man up may lay his senior out." This almost certainly describes the fate of Old Parr, who died soon after crowds had thronged to see him exhibited in the court of King Charles I. A younger man might have taken his new acclaim in stride and not succumbed, as Parr was generally assumed to have done, from being overfed and too well treated in London.

The biological paradox in question is exactly why growing older causes humans and animals to grow more fragile. There is no obvious reason that this must be so. No law of physics needs to be broken in order to design a non-aging animal. In thinking that aging is as inevitable as, say, the passage of time, we are committing what might be called the fallacy of the machine. That is, we tacitly assume that our bodies are like machines that must inevitably wear out. The flesh is assumed quite literally to be weak and doomed to eventual failure.

However, living organisms are very different from machines. The most fundamental defining character of living organisms, in fact, may be their ability to repair themselves. We don't usually die of cuts, bruises, or even broken bones. These injuries mend, and life goes on. Some animals perform remarkable feats of self-repair. Tear a starfish in half, for instance, and each half will regrow its missing half, so that you eventually end up with two healthy starfish. Do starfish live forever, then? Do they fail to age due to their extraordinary power of self-repair? We will examine that question later. For now, it might be worth thinking about why humans and most other organisms with adequate, if less than starfishlike repair capacities, fail to keep the accumulating damage of time forever at bay.

Another hint that aging is not biologically inevitable is that even organisms that *do* age don't start doing so immediately. For the first part of our lives, for instance, we actually improve in virtually every bodily function, be it physical coordination, cardiovascular strength, or immune-system responsiveness. Why we ultimately begin teetering in the other direction is the question. If we can become more and more fit early in life, why can't our physical condition keep improving throughout life?

Some cells in our bodies can achieve immortality, but only by turning cancerous. For instance, in 1951 a few such cells were removed from Henrietta Lacks, a terminally ill young woman from Baltimore. They have been irrepressibly growing and dividing in laboratory culture dishes ever since. Known as HeLa cells, they are now so numerous that they are used to study the biology of cells in hundreds of laboratories around the world. But normal cells, as opposed to cancerous cells, do not grow and divide forever. In a similar laboratory dish, normal cells from, say, skin or lungs will grow and divide for a time, then stop. So why can't "normal" cells have the immortal properties of "abnormal" cancer cells?

Another puzzling question about aging is that if animals must deteriorate and die for some reason, why do they do it at so many different speeds? Nothing more poignantly illustrates the different aging rates between humans and many other animals than the difference between our lives and the lives of our pets. I remember quite well my first dog, Spot. Spot was a puppy when I was a toddler, and a swaggeringly mature dog-on-the-make by the time I began kindergarten. When a decade later my adolescent hormones had just begun to flow, Spot had become a blind, doddering and drooling, weak-sphinctered old codger. He died before I went away to college.

Spot's life course was pretty much the same as what we humans endure. He simply sped through it in a decade and a half rather than three score and ten years. If Spot had been a mouse instead of a mutt, he would have passed through all the same stages in a couple of years. And if he'd been a tortoise, he might be looking at me right now in a quietly pitying way, wondering why I have been failing so fast.

So what does it all mean? If different animals are stuck at their particular aging rates, if the reports of exceptional human longevity have all turned out to be false, does this also mean that all the so-called antiaging drugs, diets, attitudes, and ointments are similarly bogus—the product of wishful thinking rather than scientific knowledge? What about antioxidants or exercise or vegetarianism? Even if current news on the antiaging front isn't promising when examined with a coldly skeptical eye, aren't there new therapies to combat aging in the research pipeline? If the paradox of aging is unresolved—or worse, unresolvable—does this mean that we are forever condemned to the biblical three score and ten years, give or take a decade, thanks to antibiotics? In the course of this book, each of these questions will be addressed and answered. But in order to examine and understand the paradox of aging, we must first define exactly what it is that we are talking about.

How Can We Measure Aging?

In order to examine the causes of aging, we must be able to define and measure it. As Lord Kelvin once rather pompously declared, "Until you have measured something, you don't know what you are talking about." Of course, Lord Kelvin is now famous for having deduced from the temperature of the earth that it could be no more than a few hundred thousand years old, thus demonstrating that even if you *have* measured it, you *still* may not know what you are talking about. (The earth is now known to be more than 4.5 billion years old.[3]) Nevertheless, it is always useful to be able to measure something.

So let's begin by simply stating that aging—or, as scientists sometimes call it, "senescence"—is the progressive deterioration of virtually every bodily function over time. That's simple enough, but how is it best measured? Most people assume that you can measure the speed of aging by merely recording how long something lives. It appears obvious that if something ages rapidly, it won't live very long, and if it ages slowly, then it will have a long life. If dogs live 10 to 20 years and horses live 30 to 50 years, then horses must age more slowly than dogs. But just using life span to measure the rate of aging is tricky.

To understand why, imagine, as did Peter Medawar in his essay "An Unsolved Problem of Biology,"[4] the fate of a population of test tubes in a laboratory full of busy scientists. A basic fact of laboratory life, of course, is that test tubes will be broken and replaced with new ones. If each test tube had its date of manufacture etched on it, we could easily calculate the average life span of a test tube in any particular laboratory and compare test-tube longevity in different labs. We would without question find some differences, too. But because test tubes do not deteriorate over time (if we ignore the accumulated nicks and scratches that might increase their fragility), we wouldn't be measuring anything to do with aging at all. We would only be comparing what might be thought of as the relative hostility of their environments.

What this simple example demonstrates is that things, whether they be test tubes, automobiles, computers, or humans, are destroyed for a combination of reasons—their fragility, the harshness of the environment in which they live, bad luck, and, if they age, the rate at which their fragility increases with time. Just looking at longevity by itself lumps all these causes together. So when comparing longevity among animal species or humans, we must remember that differences may be due to factors other than aging per se. The longevity of a bird in the bush, in other words, cannot be meaningfully compared with that of a bird in the zoo.

The limits of longevity as a measure of aging are also nicely demonstrated by the following. In this century in the United States, average longevity has increased from 48 years to more than 75 years. Think about photographs of people from early in the century—your grandparents, perhaps, or famous public figures. Does it seem likely that we now age nearly half as fast as people did then? Does the 46-year-old new president Teddy Roosevelt look any different from a 46-year-old now? More specifically, does an 80-year-old today look like a 40-year-old did at the turn of the century? What has changed, of course, in the past 90-odd years is our ability to combat infectious disease, the quality of our water, the sanitary conditions in which our food is stored and prepared, and a host of other public-health measures. As we will see, there is precious little evidence that we have actually slowed the pace of aging itself during the past 90 years.

U.S. Presidents Theodore Roosevelt and Bill Clinton at age 46. Even though life expectancy during Roosevelt's time at the turn of the twentieth century was only 48 years, compared with today's 75-plus years, this change is largely due to fewer premature deaths from infectious disease rather than to a change in aging rate. (Courtesy of Sagamore Hill National Historic Site/National Park Service and The White House Photo Office)

If longevity by itself is a crude tool to measure aging, is there something better? What about using some measure of physical fitness, such as speed in a 100-meter run, throughout life? Unfortunately, measures of fitness are hard to relate directly to aging. For instance, a friend of my father—"Society" Jack Sackett, as he liked to be known—was a nationally recognized sprint champion in high school and a champion debauchee thereafter. He undoubtedly achieved his peak sprint speed in high school, the last period of his life in which he seriously trained, and general dereliction probably hastened its decline as he grew older. On the other hand, I probably reached my peak sprint speed at some point in my early twenties during a fear-aided dash from the police during an antiwar demonstration. I like to think that my sprint speed has declined more slowly than Society Jack's because I have put considerable energy over the years into keeping fit, but then, I began declining from a lower plateau. I also have to admit that when I was a cocky undergraduate at UCLA, and Society Jack was laboring through his bibulous sixties, I still refused to race him 20 yards for money. I wasn't at all sure that booze, time, and degeneracy had compensated for the genetic difference between us.

Problems also arise when measuring most other physical characteristics. For instance, our ability to unconsciously control our body temperature declines as we get older, but it does so almost imperceptibly and is not noticeably impaired until relatively late in life. That is when we decide to move to Florida. On the other hand, our ability to remember new vocabulary words declines rapidly, starting in our mid-teens. As a friend of mine, who decided in his mid-thirties to learn ancient Greek, said, "Languages are a young person's game." My friend, a professor of English at Harvard, was no dummy, but he felt clearly disadvantaged in a class of eager 18-year-olds.

A different kind of problem arises if we look to another possible characteristic to measure—fertility. The main problem with measuring fertility is that there is a difference between the sexes. Women in industrialized countries, for instance, are most fertile in their early to mid-twenties, with a relatively rapid decline thereafter. Having children in the late thirties is much riskier for both woman and child than it would have been even a decade earlier, and without heroic new medical technology, women cease reproducing entirely after the ages of 45 to 50. Men, on the other hand, show no abrupt decline in fertility as they grow older. They do show a steady decline, on average, in the proportion of normal sperm that they produce. The swimming speed of those sperm also decreases somewhat, and there is a steady increase in the rate of impotence throughout life. However, all these changes are very gradual, and there are reliable reports of men siring children at age 95.

Does this mean that men age more slowly than women? As John Wayne might have said, "Not hardly, pilgrim." Women live consistently longer than men throughout the world, and they die at lower rates than men both before and after menopause.

Is there no simple, broadly suitable way to measure aging then? Can we not say exactly when it begins, how rapidly it proceeds? If not, then is Lord Kelvin right: Might we really not know what we are talking about? Further, if we can't measure aging precisely, how can we evaluate the claims of self-proclaimed antiaging medications such as antioxidant vitamins?

Despite the fact that it's difficult to measure aging in any given individual, it is possible to measure the rate at which a whole population of people, such as the population of the United States, is aging. Looking at such a population, we can actually identify when aging begins. As it turns out, aging doesn't begin at birth, as some people claim; it doesn't begin at conception, or upon graduation from college, or when we land our first high-pressure job. Aging begins at about 10 or 11 years of age, or just before we reach puberty.

How can I say this? For a whole population, changes in the probability of death at any age is a pretty reliable guide to physical condition. This probability changes in a consistent fashion in human populations and across many animal species, especially if they live in a protected environment, such as any modernized country or a good zoo. The general pattern is that the probability of death is initially high around birth, gradually declines to some low point, and then begins increasing. It increases at an accelerating rate throughout the remainder of life. The rate at which the probability of death increases as we grow older is a superb measure of how fast we are aging, in general.

We can, of course, die at any age. Eternal youth does not mean eternal life. Accidents always happen, just as they did to our test tubes. If you live long enough, even if your body does not age, you will die in a car or plane crash or be crushed by a falling safe or by eating a contaminated hamburger, or from some other misfortune. The point is that as we grow older, the probability that we will die increases, and this increase is due to the deterioration of aging.

To illustrate the basic pattern of human aging, consider women in the United States (a group for which we have excellent information). In their first year of life, they have about one chance in 1,000 of dying, but this probability declines to about one-fourth that level by age 10. Then life starts becoming riskier again. Death probabilities begin increasing by age 12, and they increase at an accelerating rate from then on. By her early thirties, a woman is as likely to die as a newborn infant, and thereafter she grows progressively more likely to die. So it makes sense to say that aging begins at the point at which the probability of death is at its minimum—or, in other words, from the point at which the probability of dying forever after increases. So in the United States, aging begins at age 10 or 11.

What if we were somehow preserved forever in this bloom of 10- or 11-year-old health and vigor? In that case, we could expect to live for about 1,200 years, on average, but about one person in 1,000 would live to be 10,000 years old. Those geezers would have been born near the end of the last Ice Age and could give us firsthand accounts of the extinction of the mastodons.

Another point worth reemphasizing is that the rate of change in our probability of death accelerates through time. It changes geometrically. That is, instead of increasing by some fixed amount each year, the probability changes by a fixed multiple. The progression, in other words, isn't from 2 to 4 to 6 to 8; it is from 2 to 4 to 8 to 16, and so on. You can get into very high numbers in a hurry using this sort of geometric progression, as any crapshooter who has tried doubling his bet each time he has lost soon discovers.

Using this mathematical progression, a useful way of comparing aging among humans and animals has been devised by Caleb Finch, a neurobiologist from the University of Southern California. Finch, who is as long, lanky, and severe-looking as any Scottish Calvinist, is anything but severe or Calvinistic. Since his student days at Rockefeller University, for instance, he has from time to time dressed up in bib overalls and straw hat and played a raucously traditional Appalachian fiddle in the Iron Mountain String Band, a group made up originally of eccentric graduate students but now made up of eccentric eminent professors. He also has a sly wit, a booming laugh, and a healthily hedonistic appreciation of fine art, wine, and spirited conversation. In addition to all that, he is one of our most knowledgeable and deepest thinkers about the aging process.

Finch likes to stroke his beard, purse his lips, and, while frowning downward as though trying to peer through his beard, pop out with some new thought that changes forever the way you look at things. He describes aging as the change in mortality-doubling time—that is, the time it takes for the probability of death to double. For modern humans in industrialized countries, the mortality-doubling time is about eight years (although it varies from about seven to 10 years, depending on the country). So a 35-year-old is twice as likely to die at age 43, four times as likely to die at age 51, and eight times as likely at age 59—a trend that hasn't escaped the life-insurance companies. Check your rates. Life-insurance rates double about every eight years and a fraction.

Animal aging can be similarly measured. A mouse's mortality-doubling time, for comparison, is about three months, a fruit fly's about 10 days.

What is most astonishing is that Finch and his colleagues have noted the same eight-year mortality-doubling time of American women in 1980, in Australian civilians during World War II, and even in Australian prisoners of war in Japanese-run camps on Java during the same period, even though the harsh living conditions in the prison camps made the overall yearly death rate some 10 times higher than it was back home.[5] I've located other evidence, which will be discussed later, that Stone Age humans, who died at more than 150 times the rate we do now, also doubled their mortality rate

11

over about the same number of years. Thus, this rate of physical decay may be as fundamental to human life as the possession of a large brain, an opposable thumb, and an unfair system of taxation.

A similar pattern of mortality-rate change also occurs in modern men, but with some interesting variations on the theme. First, as mentioned earlier, men have a higher probability of death at any age than do women. In fact, males die at a higher rate all the way back to the time of conception, as shown by the fact that many more male than female fetuses are spontaneously aborted. In fact, for reasons that aren't clear, the ratio of the sexes at conception is estimated to be nearly two males per female, whereas by birth the sexes appear in nearly equal numbers. Men are apparently just more fragile—a weaker vessel, one might say, even in the protected environment of the uterus.

Another peculiarity of men, at least in modernized countries, is that they exhibit a pronounced spike in their death rate beginning after puberty. That is, their death rate increases dramatically (tenfold!) between ages 11 and 23, then gradually declines for about a decade before settling into an eight-year mortality-rate doubling for the rest of their lives. I call this period of male life, between 12 and 23, the time of testosterone dementia, because it is a behavioral, not a physiological, phenomenon. During those years, more than two-thirds of male deaths are due to accidents and suicides, and men (or, more accurately, maturing boys) are more than three times as likely to die as women. Without accidents and suicides, men's death rates increase gradually from age 11, just as females' death rates do. Testosterone dementia makes maturing males enthusiastic warriors, hunters of large game, or fighter pilots, but poor insurance risks, as the driving records so graphically show.

Note that at an age when testosterone dementia is only a fond memory and a source of barroom yarns, men endure the same eight-year mortality-doubling time as women; however, they still die about twice as frequently as women at any particular age. Thus, women live longer than men not because they age more slowly, but because they are less intrinsically fragile.

So aging and longevity are not exactly the same. In fact, for some animals at least, it is easy to increase longevity, but it's hard to retard aging.

Try an experiment: Put some house flies in the refrigerator. If you have chosen your temperature correctly and provide sufficient food, they might live as much as 10 times as long as the flies that normally buzz around your cake in the kitchen. However, the mortality-rate doubling of the chilled flies will be the same as that of your kitchen flies, if you can overcome the urge to swat them into oblivion long enough to complete the experiment. Chilled flies simply die at a lower rate throughout life. So even though you have extended the fly's life, you haven't retarded its aging.

The reason chilled flies live longer is no mystery. You have slowed down all their physiological processes, because all biochemical processes are slowed by cold, and you have therefore created a more salubrious environment for the flies in terms of longevity. You haven't done much for its sex life, however. Flies, so chilled, do not reproduce, which is why evolution has failed to favor house flies that migrate to the Arctic for reproduction.

The same phenomenon occurs in some mammals, such as those that hibernate. Some years ago, Charles Lyman and his colleagues at Harvard University exposed Turkish hamsters to the cold and recorded differences in their response.[6] Some hamsters slipped quickly and easily into hibernation and lived considerably longer than more stubborn hamsters that tended to shiver but remain awake. However, the rate of mortality doubling remained exactly the same whether they hibernated or not.

From a medical perspective, it doesn't matter whether we increase longevity by decreasing fragility, increasing environmental quality, or retarding the aging rate. Long life is long life, assuming that one is healthy. However, health is an issue that cannot be ignored, because few of us would want to keep living interminably while growing more and more feeble.

But our medical advances to date have not increased the health of the elderly as much as that of younger people. We have increased life expectancy largely by conquering diseases of childhood and midlife. Even these successes have been largely related to improving the quality of our environment, developing antibiotics, and so on.

It's instructive in this regard to consider, for instance, just how far we have come since the turn of the century in keeping women alive in their childbearing years. Between the ages of 15 and 35, women are only about one-tenth as likely to die now as they were in 1900. This huge change has come about mainly because of increased attention to hygiene during the birth process and the availability of antibiotics to combat puerperal fever, a bacterial infection of the uterus that was often fatal. These sorts of advances illustrate how we nearly doubled life expectancy in less than a century. By contrast, all our modern medical advances have only halved the annual death rate of people in their sixties over the same time period—nothing to sneeze at, but a far less impressive accomplishment than the change in young women's mortality. Furthermore, much of this change is related to our ability to keep debilitated people alive longer.

And even that increase may be grinding to a halt. Consider our two leading causes of death at present: cancer and heart disease. If all cancer were eliminated tomorrow, human life would be extended only by about two years.[7] Eliminate heart disease and we would get about three to four years more. Cancer and heart disease together kill more than half of all people in industrialized societies. If we can add only six years to our lives by eliminating both these diseases, we are clearly pushing the limits of life-span increase by normal medical progress. Beating diseases back one by one will have progressively less impact in the future. The only hope for a dramatic increase in human longevity is new insight into the nature of aging itself and the development of treatments to slow the overall process. It is to illuminate the promise of such new insight using recent developments in evolutionary biology, comparative zoology, and traditional basic medical research that this book has been written.

2

Age Inflation and the Limits of Life

In the matter of prolonging human life, science has played no part whatever. Take the case history of one Bessie Singletree. At the early age of five, Bessie suddenly became six and entered school. On trolley cars her age remained at six until she was nine. When she was 11 years old, she was 12, and for the benefit of movies and railroads, she was 12 until she was 15. . . . On her 27th birthday Miss Singletree became 24. . . . At 40 she was 39 and she remained so until she was close to 50. At 50 Bessie was 40; at 60, 55. . . . [O]n her 70th birthday everyone said Grandmother Singletree was pretty chipper for an octogenarian. At 75 she had her picture in the paper as the oldest woman in the county, aged 93. Ten years later she passed away at the ripe old age of 109.

NORMAN INGERSOLL (1936)

Trying to understand aging requires, not suprisingly, knowledge about how old people actually live to be. Yet as the epigraph illustrates, people will lie about their age for a variety of reasons. Youngsters want to be older, the middle-aged would like to appear younger, and the very elderly would like to be believed to be even more elderly. Even when people don't lie, the very elderly often don't know how old they are because of lost birth records or a memory devastated by senility. So there was nothing out of the ordinary in Old Parr's exaggeration of his age. It is a particularly universal form of vanity among aged humans—one that is apparently more common in men than women for reasons that I'll leave to the psychologists. However, it is easy to understand the fundamental motivation. An 85-year-old man is just another codger, but a 130-year-old man such as, say, Charlie Smith, is a celebrity, a guru of longevity whose advice on successful living is broadcast far and wide. But systematic age exaggeration of the very oldest people means that for eras and geographic areas lacking accurate written birth and death records, evaluating individual longevity is an uncertain task. It isn't just self-reported ages that may be erroneous. Lies and erroneous ages can as easily be hammered into tombstones or entered into official documents as proffered over cocktails.

For the very oldest of old people—those older than 90 or 100—there are few places in the world where reliable information on death rates exists, even today. Accurate death rates of centenarians (people 100 years old or older) are available for some European countries, such as France and Denmark, that had compulsory birth registration by the late nineteenth century. Sweden is the platinum standard: It has had good records since 1750 and impeccable ones since 1850. Universal birth registration did not exist in the United States, however, until 1940 and still doesn't exist in much of the world. Therefore, although we may speak confidently of the number of centenarians currently alive in a few countries, similar information in the United States and most of the rest of the world is more speculative. Demographers who study the very elderly actually consider questionable any ages higher than about 80 in the United States.[8] The U.S. Census Bureau is aware of this problem and dismissed as fabrications more than 94 percent of the 106,000 cases in which people claimed to be older than 100 in the 1970 census. And as we shall see shortly, they may not have been skeptical enough. Age exaggeration is particularly unfortunate because determining whether the *maximum length of life* (presumably a measure of how long our bodies will last under the best circumstances) differs geographically or historically or with certain dietary or exercise habits or overall lifestyle could conceivably provide a wealth of information about how and why we age.

Given all this uncertainty, how could I have been so sure when I claimed that Old Parr's stated age of 152 was a fraud? The fact that Parr had no birth records whatsoever is suspect, but I'm most convinced by the fact that we now have many millions of unquestionably valid ages at death from around the world. Because there are now more people alive than ever before, and because precise birth records are available in an increasing number of countries, we have probably learned more about the limits of human longevity in the past 20 years than in all previous historical eras combined. And with all these millions upon millions of reliable records, there were until recently no verified records of any human living to even 120 years of age. However, on February 22, 1995, Jeanne Louise Calment, a woman born in Arles in southern France 13 years *before* Vincent van Gogh moved there from Paris, became the first verified 120-year-old person in human history. (As of this writing, in November 1996, she is still going strong.) There have been plenty of previous claimants to that age or greater, but they never have had documentation as extensive and valid as Madame Calment's. It is difficult to credit more extravagant but undocumented claims than Madame Calment's when you realize that even with all of today's improved health care, a person is more likely to be struck by lightning than to live to be even 110 years old, much less 120 or beyond.

If there is a secret to achieving a life of 100 years or longer, we have now discovered what it is. You simply need the good fortune to be born into a nonliterate culture, or one with sloppy record keeping, or one such as ours of exceptional gullibility.

17

Near the turn of the twentieth century, T. E. Young, the skeptical former president of the English Institute of Actuaries, published several editions of a monograph reexamining the authenticity of extreme longevity claims. In the first edition of his book, published in the 1870s, he found no verified claim of anyone living longer than 105. But by his last edition, in 1905, when he had accumulated thousands more death records, he would admit that the four oldest people known to that time were truly 108, 109, 110, and 113 years old. Not until 1995 did we have our first well-authenticated case of a 120-year-old. But a seven-year increase since the beginning of the century, given medical and public-health advances—not to mention the many millions of additional reliable records available since 1905—seems surprisingly small. If the maximum length of human life is increasing, it is doing so at a glacial pace.

Despite little evidence that maximum longevity has changed with modern medical progress, three areas of the world—isolated regions of the Caucasus, the Karakoram Mountains, and the northern Andes—have received special attention as putative Shangri-las, where living a healthy 100 years is commonplace. Each of these regions is characterized by Spartan farming, hard physical labor, a supportive social network, and, of course, poor birth records. All the areas have been visited at one time or another by scientists interested in factors leading to extremely long life. We need to bear in mind, though, that these scientific visits do not necessarily validate the longevity claims themselves. Scientists can be as gullible as anyone else. Martin Gardner, an amateur magician and professional debunker of pseudoscientific assertions of paranormal powers, says that it is a commonplace among magicians that scientists make the easiest dupes, because they believe that they are especially acute, trained observers. But take a scientist out of the laboratory, drop him in an alien milieu, and he's just another rube waiting to be fleeced by the locals.

As a scientist who has worked for a number of years in remote villages of Papua New Guinea, I speak from firsthand experience. I can't usually get reliable answers to questions such as, "Who owns this land?" and "How many days' walk is it to the next village?" The inevitable answer to the first question is, "I do," and the tacit answer to the second is, "As many days as you will continue to pay us for carrying your gear." It took me a while to discover this. I initially thought that walking three to four hours per day was all the locals ever did. I found out otherwise when some of my carriers forgot some personal items they wanted to have along. We had walked a bit more than two days when these fellows turned around, walked home and back, and caught up with us on the same day.

Dr. Alexander Leaf, a distinguished physician at Harvard Medical School and Massachusetts General Hospital who visited all three of these localities in the early 1970s, pointed out that in addition to being areas of Spartan farming, each of these locales was also characterized by "poor sanitation, infectious diseases, high infant mortality, illiteracy, and a lack of modern medical care,"[9] making the inhabitants' extreme longevity even more extraordinary. Clearly, if people do live exceptionally long lives under such harsh conditions, detailed studies of their genetics or way of life might provide clues about how human life might be extended elsewhere. Another reason for examining these claims more closely is that reports of similarly isolated long-lived groups continue to appear, and we want to understand how much credence we should give them.

The region about which we know the least is the small, extremely isolated Hunza region in Pakistan's Karakoram Mountains, near the western end of the Himalayas and Pakistan's borders with India and China. The Hunza are tall and fair compared with their neighbors and claim to be directly descended from Persian "wives" of Alexander the Great. Outsiders have typically employed Hunzas as load carriers on mountaineering expeditions and invariably mention their remarkable vigor, endurance, and good humor. The organic-farming evangelist J. I. Rodale, who widely publicized their longevity claims, called them the "happy Hunza." I became less convinced of their intrinsic charm after learning that one of their *mirs*, or chieftains, came to power during Victorian times by poisoning his father and having his two brothers pitched off a cliff. Maybe he did it with good humor, though.

In any case, initial reports were that many Hunza *men* lived to be

120 to 140 years old. More recent claims are less extreme, 110 years or so, but the mention of only men of this extreme age is suspicious, given that women greatly outnumber men by the age of 100 in all cultures with a history of reliable records. Nonetheless, claims of Hunza longevity have been publicized in a series of books since the 1920s.[10] The explorer Lowell Thomas visited them in 1957 to report on their unusual longevity, and the television personality Art Linkletter financed an expedition to Hunza territory in the early 1970s to study their living habits. Their longevity has been attributed to vigorous exercise, farming with optimally cured manure, a largely vegetarian diet, breast feeding of their young, and even the rock dust that floats down onto their crops of wheat and barley from the mountains above. One physician claimed that rats fed a Hunza diet suffered far less from disease and were more cheerful than rats fed white bread, sweet tea, and tinned meat—a so-called English diet.[11]

The only problem with these accounts is that the Hunza, like Old Parr, have no age documentation whatsoever. They have no written language. Everything we know about their longevity comes from the word of their *mir*. Further, old men are venerated to the degree that a Council of Elders routinely advises the *mir* on all important decisions. Therefore, although there is no evidence supporting or refuting their claims of longevity, the Hunza have means, motive, and opportunity for rampant age inflation.

A second region, one that has reported the most extreme ages yet, is in the remote Caucasus Mountains, formerly in the Soviet Union but now consisting of the independent states of Georgia, Armenia, and Azerbaijan. People in the Caucasus have been visited more frequently by scientists than the Hunza. Here a number of people claim to be more than 150 years old: The oldest among them was the previously mentioned Shirali Muslimov, who died at the declared age of 168 years, seven years after his picture appeared in *Life* magazine. The television show *60 Minutes* followed up more than a decade after *Life*'s story with a segment on the same village, and somewhat ironically, given the institutional skepticism of *60 Minutes*, seemed to swallow these stories whole. An especially healthy lifestyle, at least as we currently understand it, could not be responsible for the longevity reported there. The local diet contains plenty of meat, dairy products, wine, and sweets. The oldest woman Dr. Leaf interviewed on his visit to the region, supposedly 130 to 140 years old, said she had smoked a pack of cigarettes a day for more than 60 years and started each morning with a shot of vodka before breakfast.

There is no reliable documentation of these longevity claims, either. The region has been repeatedly ravaged by wars and social dislocation, especially during the late nineteenth century, when the putative centenarians would have been born. Even the normally rudimentary church records typical of this period rarely exist, and no Soviet identification documents were required before 1932. Birth dates for identification cards of people born before 1932 were determined from oral interviews.

Of course, the official statistics of the Soviet period are not widely believed in any area. What can one make, for instance, of the Soviet Union's claim from its 1959 census that it had the lowest death rate in the world?[12] Regarding the special claims of the people of the Caucasus, and of Georgia in particular, Zhores Medvedev, an emigré Russian geneticist, tells us that because Stalin was a Georgian, he enjoyed hearing stories about extremely old Georgians. Not surprisingly, local authorities were eager to satisfy his desire for these stories. Thus, in the 1959 census, Georgia provided 97 percent of all Soviet centenarians, even though it had less than 2% of the Soviet population.

At least one fraudulent centenarian from the Caucasus was exposed when he had the misfortune to have his photograph appear in the government newspaper *Izvestiya* on the occasion of his alleged 128th birthday. Soon afterward, *Izvestiya* received a letter from the man's fellow villagers revealing that he was a World War I deserter who had been using his father's papers to avoid detection. He was actually 78. Medvedev claims that similar deceptions were common throughout the former Soviet Union. Also, because most people in this area are Muslims, there is ample room for confusion in translating the 10-month Islamic calendar into our 12-month one. Ultimately, then, we must take claims about longevity in this area on trust, too, and there is little reason to suspect that our trust would be well placed.

The third region—one that has received the most thorough attention from scientists, including repeated visits over almost a decade—is the Andean village of Vilcabamba in southern Ecuador and its environs. There is a tradition of people in this region claiming to be more than 120 years old, and many more claiming to be in their 90s and early 100s. Dr. Leaf was tentatively convinced that the claims were valid when he first visited Vilcabamba. Dr. Donald Davies, a gerontologist from the Medical College of London who visited Vilcabamba in the early 1970s, was so thoroughly gullible not only about Vilcabamba but also the Caucasus and Hunza region, that one wonders whether he may have been on some unusual medication at the time. He simply reported as fact everything he was told—150 years, 160 years, no problem. His book (*The Centenarians of the Andes*) attributes the longevity of the Andeans largely to a positive attitude among the elderly and an abundance of trace minerals, such as gold, magnesium, and cadmium, in the soil.[13]

Superficially, there is little in life in the Ecuadoran Andes that might lead one to expect the people to be particularly long-lived. If anything, one might expect the reverse. Although they eat relatively little and work very hard, the Andeans also smoke and drink alcohol extravagantly. In one of the unintentionally funniest lines in his book, Dr. Davies reports, "At times of stress the males drink themselves into a stupor; they also do this regularly on weekends. The women show more signs of stress, looking much older for their years and they don't live so long."

So why did this information initially appear so convincing? For once, there were actual baptismal records in the local church and Civil Registry records as far back as 1860 which could be checked for verification. Records, any records, were more than had been available elsewhere. In addition, Ecuadoran physicians always accompanied the visiting scientists, and the physicians claimed to have checked the records personally.

And yet, when Dr. Leaf returned to Vilcabamba a second time

four years later, he found that Miguel Carpio, who had previously been the oldest man in the valley at 121 years of age, had miraculously become 11 years older. When he demanded to see Carpio's baptismal certificate, it seemed to have disappeared in a church fire. The earliest book of existing records had also had its first seven pages torn out.

The confusion about birth records was ultimately explained when it came to light that the people of Vilcabamba intermarry almost exclusively within the same mountain valley, and the same few names are used over and over. There is also a local tradition of reusing the same name within a family, so that a child born after an older sibling dies is likely to be given exactly the same name. Thus Micaela Quezada claimed to be 106, and her baptismal record, underlined emphatically by local officials, made her 104, which is not too far off. However, the names of the parents on her baptismal certificate were very different from those of her actual parents. When questioned about the names on the baptismal certificate, she said, "Oh, yes, of course: That's my cousin [*actually her aunt*]. . . . She was older than me and died 30 or 40 years ago."[14]

Several years after the initial spate of scientific visits, Richard Mazess, a radiologist, and Sylvia Forman, an anthropologist, visited Vilcabamba determined to get to the bottom of this story once and for all.[15] They thoroughly reviewed the skeletal condition of the people, looking for arthritis and osteoporosis. They performed a house-by-house census; checked all birth, death, and marriage records that they could find; and cross-checked the various documents against one another. The people whose ages they could reliably determine from records showed no differences in the degree of skeletal deterioration from similarly aged people in the United States. As they worked their way through a bewildering maze of documentation, Mazess and Forman found a consistent pattern of age inflation and a consistent pattern of inconsistency in the records. For instance, the Miguel Carpio who had magically transformed from 121 to 132 in only four years was officially recorded as having died at 112 years. In actuality, he died at 93. Like the fictional Bessie Singletree of this chapter's epigraph, when he was 61, he reported that he was 70; five years later, he was 80; and when he was really 87, he said he was 121. His mother was, in fact, born five years after his own stated birthdate, something that even the heroic modern advances in reproductive technology have not been able to duplicate.

Mazess and Forman ultimately found that none of the 23 self-proclaimed centenarians had actually reached 100 (their average age was 86), and none of the 15 "nonagenarians" had reached 90 (their average age was 81.5).

In his later evaluation of Vilcabamba's inhabitants, Dr. Leaf felt he had at least found the motive for their continually increasing age inflation. On his second visit, traveling over a newly paved road that had been rough gravel on the first go-round, he was met by the governor and a local band and hailed as the economic savior of the region. His articles in *National Geographic*, *Scientific American*, and *Nutrition Today* had brought much-appreciated attention to the area. The governor was calling the village's old people "our oil wells." Even in 1978, when the myth of Vilcabamba was being debunked thoughout the scientific community, Japanese investors were negotiating with local authorities to build a high-rise hotel, and an American entrepreneur was planning to market bottled water from Vilcabamba's stream.

It would be a mistake to assume that these occasional high-profile frauds define unique moments of age inflation. It is systematic throughout the world, appearing wherever the deficiency of records allows. Consider, for instance, the geography of your chance of living to be a centenarian. No matter where or when you were alive, you were much more likely to live to be 100 years old if you or your relatives were illiterate, or if you lived in a place frequently ravaged by infectious disease and far removed from modern medical care.

Currently, Sweden has among the longest life expectancies in the world (the third longest for men, at just over 75 years, and the fifth longest for women, at nearly 81 years) and an excellent history of record keeping for better than 150 years. In fact, comparison with Swedish data is one method used to assess the validity of mortality records from other countries. The proportion of centenarians in Sweden is about 5 per 100,000 people. Japan, with the greatest life expectancy in the world, has about the same proportion of centenarians as Sweden. But remarkably, according to "official records," you were more than twice as likely to live to 100 if you resided in turn-of-the-century Argentina, Bolivia, Brazil, Bulgaria, the Philippines, Russia, or Ireland than if you resided in present-day Sweden or Japan!

Also, in the United States as elsewhere, the number of centenarians has decreased as literacy rates have increased. For instance, in 1850, the centenarian rate among Americans was 11 people per 100,000. By 1910, it had fallen to 4 per 100,000. There was also a puzzling difference between death rates in extremely elderly African Americans versus Caucasian Americans. In 1910, African Americans reached 100 years of age at more than 20 times the rate of white Americans. This rate has been steadily falling since then (only four times as high by 1960) as literacy among blacks has risen.

This consistent relationship between literacy and long life is not necessarily due to conscious fraud. It's just that, in the absence of knowledge, people exaggerate. And such a habit is apparently responsible for another common demographic anomaly—the mortality-rate crossover, in which one group of people dies at higher rates than another group early in life, but dies at lower rates later on. For instance, perusing vital statistics compiled by the United Nations in 1990, you will find that in the small southern African country of Malawi, whose early-life death rates are so high that the country's current life expectancy is only about 40 years, older people have a *lower* risk of dying at any particular age than people of similar age in the United States and Japan. Japan is currently the country with the greatest life expectancy in the world for both men (over 76 years) and women (almost 83 years).

There are two possible explanations for this rather apparent paradox. One interesting idea is what we might call "survival of the hardiest." That is, the mortality crossover may represent selective weeding out of weaker individuals. The harsh conditions of life in Malawi could conceivably kill off all but the physically hardiest early in life, leaving only exceptionally disease- and death-resistant people surviving to greater ages.

A similar explanation has been advanced for a similarly puzzling observation in the United States. Throughout this century, life expectancy has been lower—considerably lower—for African Americans than for Caucasians. It's not surprising, then, that from birth until their late seventies, blacks have higher annual death rates than whites. What *is* surprising is that when Americans reach their late seventies, these death rates traditionally "cross over"—as do the U.S. and Malawian death rates—and at all subsequent ages, black Americans die at lower rates than whites. The traditional explanation for this pattern has been survival of the hardiest. That is, because of the acknowledged harsher social and economic conditions in which blacks live on average, only the most healthy and fit are likely to make it to their late seventies in the first place. Those who do survive are so hardy that they thrive even in their difficult living conditions.

A second, more likely explanation is that the death rates for older people in Malawi or among very elderly American blacks are simply false, due to a lack of birth records combined with age inflation. Even when old people *do* know their true age, they *still* exaggerate. Appreciating the validity, or lack of it, of longevity claims is crucial to determining whether or not aging rates have changed historically or are lower in some parts of the world.

Most demographers now agree that this "crossover" between blacks and whites in the United States has resulted from a lack of accurate birth records. A good clue that this is so is that the age of the crossover has steadily risen from 76 in 1960 to 85 in 1980 to 90 in 1987.[16] Recently, the demographer Samuel Preston appeared finally to have killed the myth of "survival of the hardiest." He pointed out that people exposed to bad health conditions generally die at higher rates throughout life—even in the very latest years. Cross-checking census and Social Security records from early in the century with ages on death certificates of several thousand black people who died in 1985 at ages greater than 65 years, he and a colleague, Irma Elo, indeed found inconsistencies in more than half of these records.

Because of rampant age inflation, professional demographers have developed a number of clever tricks for assessing the reliability of the official records for the extremely elderly. One thing they look for is "age-heaping"; that is, a trend for an unexpectedly high number of people with ages that are "round numbers," such as five or 10. In some censuses, 10 times the number of people are recorded as age 70 than as 69 or 71. This is a sure sign of shaky information.

Another sign that information is unreliable is a suspiciously low death rate at older ages. In countries with good records and low overall death rates, such as Sweden and Japan, the odds of dying after age 100 are about 50 percent per year. Calculating this figure over five years, there should only be about 3 percent as many people who are 105 years old or older as there are people who are 100 years old or older. So finding that there are 40 percent or 50 percent as many 105-year-olds as there are 100-year-olds is a sure sign of age inflation. By this criterion, very few countries have reliable statistics on their very oldest people.

An interesting and instructive exception to the general rule about the reliability of age claims and literacy is among the Han, the ethnic majority in China, representing about 95 percent of its people. Even illiterate people can usually supply their precise date of birth in this culture, because Han birth dates have astrological importance. Also, the Han calendar consists of an easily remembered cycle of "animal" years (each year will be associated with one of five different qualities of one of 12 different animals) that repeats every 60 years. By examining age-heaping and several other criteria to expose age inflation, the demographers Ansley Coale and Shaomin Li determined that in contrast to Han Chinese, other ethnic groups in China *were* prone to systematically inflating their ages. For instance, Xinjiang Province, in which the Wei minority make up almost half of the population, shows extensive age-heaping: Although the province contains only about 1 percent of the Chinese population, it accounts for 84 percent of all males claiming to be more than 110 years old.

Most of China's other "supercentenarians" are spread among other provinces with high proportions of ethnic minorities. One should bear this in mind when one comes across news items such as the front-page story in the March 19, 1990, *Wall Street Journal* reporting that an area of Guangxi Province in China had been newly discovered to abound with centenarians—a place where 90-year-olds were unexceptional, and octogenarians were comparative whippersnappers. Later in the article, one finds that most of the old people are Yaos, a polytheistic ethnic minority inhabiting mountainous parts of relatively inaccessible southern China, Thailand, and Vietnam. Sound familiar?

3

Has Aging Changed
over Time?

*If they [the Houyhnhnms] can avoid casualties, they
die only of old age. . . . [T]hey live generally to
seventy or seventy-five years. . . . [S]ome weeks
before their death they feel a gradual decay, but
without pain.*

SWIFT, *GULLIVER'S TRAVELS*

Bearing in mind that it is unwise to credit any extreme ages without firm documentation, let's now look at the history and geography of human aging and longevity. If people living uneventful
bourgeois lives in modern industrialized countries and harshly treated
prisoners-of-war both double their risk of death every eight years or
so as they grow older, how universally does this represent human
aging? Did the Greeks of Socrates' day, the sun-worshiping Egyptians
of 5,000 years ago, and our Ice Age ancestors huddling against the
night in smoky caves also age at the same rate? Might the 20-year

life expectancy of Neanderthal humans living 60,000 years ago and the 27- and 48-year life expectancies of the ancient Romans[17] and turn-of-the-century Americans, respectively, reflect only progressively less hostile environments as deaths from famine, warfare, and disease became less common? Or has the manner in which we age changed throughout history? Might the details of aging differ in different parts of the world today?

A Brief Biological History of Humanity

Modern humans appear to have migrated out of Africa about 120,000 years ago. As they spread through the rest of the world over the next 100,000 or so years, they replaced by force, guile, or interbreeding their last remaining relatives, the Neanderthals. During this Paleolithic period, people survived by hunting wild game and gathering fruits, nuts, and vegetables. When they had depleted the local food supply, these small bands—maybe no more than a few dozen people— moved on, perhaps returning later when the fruits, roots, and game had replenished themselves.

The amount of food required by a human group limited its number. Groups could not be so large as to need more food than could be hunted down or gathered within a reasonable walking distance from their temporary encampments. If groups became larger, even an area they had never previously visited would not have enough food for everyone within a reasonable hunting distance. Then tension would have arisen over who got how much of the limited food available, and the resulting quarrels and disagreements would ultimately have led to group division, with breakaway bands going their separate ways.

This pattern began to dissolve 5,000 to 10,000 years ago, when the development of agriculture and the domestication of wild animals made settling in one place for long periods possible. Nearby food was suddenly reliably abundant. It could also be stored (in bins or on the hoof) for use over the winter or during droughts. Group size was no longer limited by "natural" food abundance, and permanently occupied villages and towns began to develop. Clustered, permanent populations of hundreds or thousands were now possible, and with the increasing sophistication of agriculture, animal production, and transportation over the next few millennia, we ultimately arrived at the crowded megalopolises of today.

In terms of health (and perhaps longevity), the development of agriculture was not an unalloyed benefit. Farming decreased dietary diversity, for one thing, as people were no longer forced to scrounge for every available food resource. Because no single food provides a full complement of vitamins and nutrients, reducing the diversity of foods eaten may have led to certain nutritional deficiencies. One reason to suspect that this was the case is that with the coming of agriculture, people's stature dropped dramatically—by as much as six inches—as we know happens when people suffer nutritional deficiencies during childhood and adolescent growth.[18] In fact, adult human stature has once again begun to approach that of late Paleolithic times only during this century in well-fed countries.

The development of agriculture also resulted in new opportunities for infectious diseases (as contrasted with genetic diseases such as cystic fibrosis and Down syndrome, or degenerative diseases, such as cancer, atherosclerosis, and Alzheimer's disease—that is, the diseases of aging). So far as we can tell, diseases such as smallpox, measles, cholera, and tuberculosis—responsible for so much death and misery in historical times—troubled humans consistently only after the advent of agriculture, which brought about the development of sizable towns and cities.[19] Such diseases require sizable populations to persist. Infectious diseases, like wildfire, require fresh fuel to stay alive. From the viewpoint of an infectious disease, fresh fuel is someone who has never had the disease. If someone catches a disease, he or she ultimately dies or recovers, and after recovery is then generally immune to a recurrence of that disease. This is our so-called acquired immunity, the part of the immune system destroyed by AIDS.

Even diseases we seem to get repeatedly, such as colds and flus, behave this way. When we get subsequent infections, they are due to different viruses—that is, viruses of a different origin or those that have mutated beyond recognition by the immune system. Thus, to an infectious disease, any population is composed of people who are

31

either sick (already infected and contagious), recovered (nonconta-
gious and immune to further infection), or as yet unexposed (susceptible).

If the rate of new infections is higher than the rate of disappear-
ing infections, a disease will live on. If there are too few newly available susceptibles (such as new babies), the disease will rip through a
population, killing a certain initial number, and disappear. All survivors will be immune, and the disease will have no place to live. With
no new susceptibles to attack, it will die out.

Depending on the particular traits of any disease—how deadly,
how contagious, how long infections typically last, and so on—there
will be a critical minimum number of people required to sustain it.
For instance, the measles require a population of about 300,000
people. This means that in cities smaller than 300,000 people, the
measles are constantly going extinct, only reappearing when infected
outsiders bring it back into the community.

A major, somewhat fortunate consequence of a disease's requirement for fresh fuel is that immunization can entirely eradicate diseases even when some people are not immunized. Immunization needs
to decrease the susceptible number of people only below the critical
number. Therefore, smallpox was eradicated worldwide by a 12-year
intensive vaccination campaign that began in the late 1960s. When
the campaign started, 10 million to 15 million people per year worldwide caught smallpox, and about 2 million of them died. By 1979,
the smallpox virus survived only in a few high-security medical laboratories. As it turned out, vaccinating 70 to 80 percent of susceptible
children—which was possible even in nonindustrialized countries—
was sufficient to wipe out the disease.

In medically sophisticated parts of the world today, with *most*
serious infectious diseases under control, we tend to forget about the
impact of such diseases in the past. But until antibiotics became
generally available in the 1940s, infectious diseases were by far the
most common cause of death. In the United States in 1900, for instance, more than twice as many people died of pneumonia, flu, and
tuberculosis than died of today's "Big Three"—heart disease, cancer,
and stroke. Today, pneumonia, flu, and tuberculosis combined kill at
least 10 times *fewer* people than the Big Three.

So as human populations grew and clustered in increasingly larger villages and towns, the population threshold for more and more diseases was reached. Simultaneously, new ways appeared for serious diseases to arise. People began living in close contact with domesticated animals—cows, horses, sheep, pigs—and some diseases (viruses, in particular), are prone to jump from one species to another, given sufficient time and contact. Thus, human smallpox probably jumped from cattle, the most virulent form of malaria from birds, and a number of flus from pigs. Cities and towns supported by local agriculture, therefore, were fertile grounds for the birth and spread of human diseases.

As the millennia passed, a few subtle inroads were made against infectious diseases. Some awareness of public health developed, and as people linked conditions such as fouled water with outbreaks of disease, attempts were made to get rid of waste without fouling drinking water. But these measures were limited in scope and not sufficiently emphasized until the late nineteenth century, when Louis Pasteur proved that small organisms such as bacteria and viruses—not bad air, wretched smells or tastes, or an imbalance of internal fluids—caused infectious diseases. Shortly thereafter, antibiotics were discovered, and by the middle third of this century were generally available in most of the industrialized world. Infectious disease would never again be the chief killer of humans. The expectation of life soared.

Are We Aging More Slowly Now?

How does this biological history of humanity relate to the history of aging and longevity? If life expectancy is now greater than ever before, does that mean we are aging more slowly than ever before? Have our bodies materially changed in some way in the past few thousand years? Does our lifestyle today accelerate or retard aging as it occurred in our distant past?

One bit of evidence that suggests we are not aging differently these days is that the age at which a person is considered "elderly" hasn't seemed to change over time. We find no historical accounts of 40-year-old men who were considered elderly, even though in, say, ancient Greece (and all other known cultures until the eighteenth century) life expectancy was no greater than about 30 years. Being considered elderly at 40 seems a fate reserved for professional athletes and models, even in ancient times. The only difference is that now the psychic pain of being old at 40 is often eased a bit by the attendant economic rewards. So Alexander the Great, for instance, was considered to have died young, although at 33 he was older than the life expectancy for his time. On the other hand, Plato and Sophocles were considered to be old when they died at about age 80 and 90, respectively, and it was considered noteworthy and amazing in the second century B.C. that the Roman statesman Cato, the Elder, would begin to learn Greek at the age of 80. (How well he spoke it at 85, when he died, is a different issue.)

The reason for the discrepancy between ages considered "old" and these ancient life expectancies is that life expectancy is just the average of everyone's age at death. If there are a lot of infant deaths, the large number of very low ages will reduce the average regardless of how long adults typically live.

I can't imagine a more vivid testament to the impact of modern standards of hygiene and medicine than to note that in industrialized countries today only about 1 percent of babies die before their fifth birthday. Compare this with medically undeveloped cultures such as the primitively agricultural Yanomamö of the Brazilian rain forest, or even Africa as a whole in the 1960s, where as many as half—half!—of all children died before they turned five.[20] I wonder that in such cultures people are not too depressed to continue having children.

As a consequence of all these infant and childhood deaths, historical life expectancies can be very misleading. For instance, if in a hypothetical population half of all babies die before their fifth birthday but everyone else lives to age 90, life expectancy would be the late forties. And in that case, life expectancy would severely misrepresent the length of a typical adult life and could therefore give us no idea of how aging may have changed over time.

The rate of infant and childhood mortality in medically naive cultures of the past has usually been underestimated, falsely increasing estimates of life expectancy at those times. The reason for those distortions is that what we know about longevity in the distant past is generally inferred from reading tombstones or estimating the age at death of skeletons excavated by archeologists. In many cultures, however, the death of young infants is not recorded, and normal burial and burial rites are not observed. Moreover, the composition of infant bones makes them less likely to survive millennia of burial in a recognizable form. By failing to compensate sufficiently for this factor, the ecologist Edward Deevey presented a number of erroneously high life expectancies (in the thirties for classical Greece and Rome; nearly 50 for medieval Europe) for ancient times in a famous paper in *Scientific American*.[21] Everyone who has seriously studied the matter since then agrees that Deevey's estimates of life span are too high, although they seem to be the ones most often repeated in textbooks and the popular press.

The major pattern in the rate of aging that emerges from our best estimates of life expectancy is that from very ancient times until the eighteenth century, when some appreciation of hygiene and public health developed, little changed and life expectancy remained lower than 30 years. Somewhat surprisingly, there is no evidence of a decline in longevity as infectious diseases became more common with the development of cities and towns. This might be explained by the fact that the growth of cities brought more reliable food sources, which compensated to some extent for the rise of infectious disease. And once public-health measures began to be taken with sufficient seriousness (around the turn of the century), followed by the development of antibiotics, childhood mortality plunged, and life expectancy increased in proportion.

Also clear is the fact that because of massive early mortality in premodern cultures, changes over time in the life expectancy itself fail to reveal much, if anything, about the typical longevity of adults, much less the timing of growing old. After all, despite these 20-something life expectancies, famous people such as Plato and Sophocles lived lives that would be considered lengthy even by modern standards. But in general, the evidence suggests that for most of human history, most adults have lived only to be 30 to 40 years of age. Life was short—remarkably short. A few lucky individuals may have lived to be what we would consider elderly today, but they were clearly exceptional. Was a 50-year-old truly a codger in ancient times?

One of the best sources we have for understanding aging in preliterate times comes from a remarkable archeological excavation near the shores of Lake Erie in northern Ohio in the United States—the so-called Libben Site. In 1967–1968, more than 1,300 skeletons were exhumed from ancient cemeteries there. Between 800 and 1100 A.D., these people hunted and gathered on the edge of a great swamp bordering the lake. Perhaps toward the end of this period, they also cultivated a little corn. The skeletons were exceptionally well preserved and were excavated with exceptional care. Estimates put the ages of death from prenatal babies to more than 70 years.

Assuming these estimates are roughly accurate for the Libben people, life was apparently nasty, brutish, and short. The life expectancy of an average adult was only about 34 years. Although a few people did reach respectably old ages by today's standards, they were very few indeed. In the Libben community, even 15-year-olds had only about a 5 percent chance of reaching age 50, much less 70. A 50-year-old was indeed a rarity; a 70-year-old may have been the Jeanne Calment of that time.

But all evidence suggests that these short lives were due to the harshness of the environment, not a biological difference in the rate that people deteriorated over time. The rare 50-year-old codger would not have been as decrepit as a modern 80-year-old—just exceptionally lucky for his time. We suspect this because some of the oldest written records of human history—much older than the Libben Site—present aging pretty much as it is today. The nearly 5,000-year-old figure of a bent osteoporotic man leaning on his staff, for instance, means "old age" or "to grow old" in Egyptian hieroglyphics. Ancient Egyptian texts indeed describe most of the medical conditions associated with aging today—heart pain and palpitations, tumors, deafness, cataracts, incontinence, and constipation. The Egyptians also considered the practical limit of human life to be 110 years[22]—an age that is reached even today in the countries with the greatest life expectancies by only about one person in 10 million. The Egyptian Pharaoh Pepi II is claimed to have reigned for at least 90 years more than 4,000 years ago, though his age at assuming the throne is unknown. Although documentation on the length of Pepi's reign is weak, the Cambridge Egyptologist John Baines assures me that Ramses II, a Pharaoh who reigned more than 3,000 years ago, had a well-documented reign of 67 years and was clearly not young when he came to the throne following the death of his elder brother. Ramses was likely at least 90 and perhaps as old as 100 when he died.

Of course, nearly all we know about the very old in the past comes from accounts of the lives of famous people—royalty or famous artists. The longevity of kings and emperors is particularly well documented because succession from one monarch to another was considered a particularly noteworthy event, and because genealogical records of royal families were meticulously maintained. Naturally, there is no reason to think that the longevity of royalty was representative of the rest of society. Princes and potentates were not likely to be malnourished or to die from overwork, and perhaps for this reason they lived to surprisingly modern ages for as far back in time as we have good records. In addition to the longevity of pharaohs, we know that among the rare Roman emperors who outwitted their enemies long enough to die nonviolently, several lived into their late seventies. The first six kings of England to die natural deaths (ruling between 1066 and 1400 A.D.) all lived to ages between 56 and 68. So for at least as long as we have written records, it was possible for people to live to ages that we still consider elderly today.

If there was a time in which people never lived to modern ages, it may have been in the very distant past—tens of thousands of years ago—when modern humans coexisted with Neanderthals. Of course, no written records from the late Ice Ages exist, but it is possible to estimate the age of death of skeletal remains using a wide variety of skull and bone characteristics (closing of growth plates, condition of long bone ends, extent of tooth wear, etc.). Since the first Neanderthal skeletons were discovered in 1856, more than 150 skeletons of Neanderthal adults (plus about 100 children) have been unearthed in the Middle East and Europe. According to skeletal estimates, none of the adults appears to have lived beyond the early forties.[23]

We can be certain that Neanderthal life was no picnic. One particularly well-preserved skeleton is of a man who apparently died in his late thirties. Healed injuries show that, at that age, he was blind in one eye, had a withered right arm, and walked with difficulty due to foot and leg injuries. On the other hand, if he survived as long as he did with those injuries, he would have had to rely on the care and kindness of others. Such attributes do not normally spring to mind when thinking of Neanderthals. One hundred fifty skeletons is not really a big enough number to conclude with any certainty that living to 60 or 70 was impossible 50,000 years ago, but it may have been.

So does this mean that the aging rate, or the speed of physical deterioration, is the same today as it has been at least since Neanderthal times, and that we are longer-lived simply because of a safer environment and better sanitation and health care? If there is a gold standard of measuring aging, something that doesn't depend so much on environmental harshness, it is mortality-doubling time, not life expectancy or maximum longevity. Remember my earlier generalization about humans aging versus that of fruit flies and mice: Humans age more slowly because the amount of time it takes for their probability of death to double is much greater. This doubling time is about eight years for modern humans versus 10 days for fruit flies and three months for mice living in climate-controlled laboratories. One way to think of these differences is to imagine that fruit flies age about 300 times as fast as humans (eight years is roughly 300 times as long as 10 days) and mice 30 times as fast. So, how has the mortality-doubling time changed over the millennia, and how variable is it today?

Humans, unlike mice and fruit flies, are not frequently found living in climate-controlled laboratories. We live in a complex world with various unpredictable dangers, such as wars, viruses, and in-laws. Yet for all this environmental complexity, our mortality-doubling time is pretty constant, varying throughout the world and throughout history less than threefold, from about seven years to about 26 years. The larger numbers, which seem to suggest slower aging, suggest something quite different. They all come from countries with low life expectancy and, in actuality, represent special situations in which young people die at unusually high rates rather than the elderly dying at low rates. If we consider death rates only from age 40 on, mortality doubling is much more constant, or if we eliminate deaths due to accidents and violence, as the demographer Jay Olshansky of the University of Chicago has done, mortality doubling stabilizes nicely in the seven- to ten-year range.

The historical perspective is similar. As overall death rates have dipped during this century and life has lengthened, the mortality-rate doubling time has decreased, because we have made more progress combating causes of death in early life than in later life. Today's longevity champions, the Japanese, averaged only 43 years of life at the turn of the century, and their mortality doubled every 16 years. Now they average nearly 80 years of life, but their mortality rate doubles in only eight years. The same general pattern is repeated in all countries in which such information has been collected.

Even in preliterate, protoagricultural cultures such as that of the fierce Yanomamö in the Brazilian Amazon forest, with a 15-year life expectancy at birth and a 40-year life expectancy in adulthood, there is a 19-year mortality-doubling period similar to that of modern Afghanistan. For the primitive Americans hunting and gathering 1,000 years ago at the Libben Site, who could expect to live to only about 34 years if they made it to adulthood at all, the risk of death doubled in a reasonably modern 11 years, about the same as in modern Colombia and Bangladesh. We can't ignore, however, that life in ancient America was dramatically harsher than that in any modern country. The odds of dying at any particular age were some 100 times higher at the Libben Site than in modern Bangladesh, which is worth remembering when we fantasize about a simple, carefree life in an imaginary Edenic past.

So although we have succeeded remarkably well over the millennia in making our environment safer, we seem to have been unable to affect the rate at which our bodies deteriorate.

4

Is Aging Genetic?

*Come for your inheritance and you may have to pay
for the funeral.*

YIDDISH PROVERB

For all the millions upon millions of mice that have grown old and died in medical laboratories around the world over the past half century, none has lived longer than five years, and certainly none has approached the life span of humans living in even the harshest conditions. So the legacy of having mice for parents is that you will live fast and die younger than you would if your parents were people. In this trivial but revealing sense, aging is indeed genetic. But more relevantly, if aging is to some extent genetic, might my aging genes differ from yours? Do some families age more quickly than others? And if, as a species, we are deteriorating today at about the same rate that we have for the past 10,000 years, doubling our chances of dying every seven-to-ten years like clockwork, does this mean that our aging rate is genetically fixed and unchangeable?

Jeanne Calment, the French woman who in 1995 became the first authenticated 120-year-old human in history, had a brother who died when he was 97 years old. Her family has had a local reputation in western Provence for exceptional longevity over the past several centuries. We have probably all known families in which everyone seemed long-lived. A former girlfriend of mine came from such a family. Her grandfather died in bed at age 96 while sipping a Coke that we had convinced him in his dotage was a beer. His brother committed suicide at 92. At age 65, my girlfriend's mother looked 45 and worked as hard as a 25-year-old. Everyone assumed that once my girlfriend was rid of me, her family history ensured that she could expect to lead an enjoyable and lengthy life, and so far she has.

These anecdotes suggest that long life does run in families. Or, to put it another way, there are genes that affect aging, or at least the development of late-life diseases and, therefore, longevity. If so, discovering the nature of these genes could tell us a lot about how we age and potentially guide the search for therapies for aging—the pharmaceutical fountain of youth. But these stories are just that—stories. What about hard information?

Geneticists are usually quick to point out that the effects of nature (that is, genes) and nurture (that is, environment) cannot be considered separately, but too frequently this message fails to hit home with nongeneticists. This point is particularly germane to understanding how genes might influence aging and its diseases.

For instance, piles of evidence suggest that certain genes have a major impact on the development of atherosclerosis, probably the major disease of aging in the Western world. One of those genes is the Apolipoprotein E, usually abbreviated ApoE, which is involved in processing dietary fat. People with one form of the gene, called ε4, have higher blood cholesterol (as well as higher LDL, or "bad" cholesterol) levels than people with other forms of the gene. Finns have the highest rate of atherosclerosis in the world and also have one of the world's highest frequencies of ε4. The Japanese have the world's lowest national rate of atherosclerosis and also among the world's lowest frequency of ε4.[24] So you could call ε4 an atherosclerosis gene. But this would be misleading, because the world's highest frequency of ε4 is found in a country—Papua New Guinea—where until recently atherosclerosis was virtually unknown.

People living in the bush in Papua New Guinea eat a low-fat diet (less than 5 percent fat, compared with 30 to 40 percent fat in an American diet) from necessity rather than choice.[25] Their daily life also involves exercise at levels that would cripple or kill most Americans, even the athletically inclined. A friend of mine who has run dozens of marathons told me that a two-day walk he took in the mountains of Papua New Guinea—routine among the locals—was the hardest thing he had ever done. So it's not surprising that Papua New Guineans don't get atherosclerosis, even those whom malaria, tuberculosis, or pneumonia fail to kill off so that they live until their sixties or seventies.

But when the Papua New Guinean environment changes, those ε4 genes do take effect. One man from a village in which I've worked was lucky enough to be hired by a mining company to learn to operate heavy excavation equipment. He moved to the mining town and began eating a fine Western diet of beef and gravy. He loved every minute of it—and died of a heart attack at 45 years of age, before his mining career ever got under way. Now his home village is suing the mining company for damages. They say they never had anyone die of a heart attack before.

So genes operate not in a vacuum but in a specific environment. This is something to bear in mind when reading of the discovery of new "longevity" genes. For instance, there is another form of the ApoE gene, ε2, which appears to lower blood cholesterol and therefore probably protects against developing atherosclerosis. Is this a longevity gene? It depends on the environment. Where people eat a lot of fat and don't exercise, it may well be a longevity gene. In fact, French centenarians are about twice as likely to have this gene as the French population as a whole.[26] But in other environments, the gene may well have little or no effect. Similarly, within a few years, it is possible that people with a common hereditary heart defect called mitral valve prolapse will live longer than people without it. People with this single genetic defect often have no symptoms, and even those with symptoms usually have fairly minor ones. However, a secondary effect of mitral valve prolapse is to reduce blood pressure and body weight, both conditions that are associated with longer life in other circumstances. Ultimately, we may be so successful at replacing the most seriously defective mitral valves, regulating the lifestyle of those affected, and determining the best time of life to replace the valves that people with mitral valve prolapse will outlive those without it due to the condition's effects on blood pressure and body weight. People with mitral valve prolapse would probably not thrive in the Papua New Guinean bush, but they just might live longer than average in a culture of extreme medical sophistication. Does this mean that the gene that causes mitral valve prolapse should be considered an antiaging or pro-aging gene?

What these examples suggest, besides the difficulty in defining genes with respect to longevity, is that unless we understand how a particular gene is influenced by a particular environment, it will be difficult to translate the effects of genes from animals to humans. This is why most gerontologists are hesitant to claim too much about the relevance to humans of genes now being found with increasing frequency in simple organisms such as fungi and worms that seem to slow aging dramatically. It is difficult to draw parallels between human and worm and fungal environments.

So keeping these caveats in mind, let's consider the heredity of aging and longevity in people. We know that there are genetic predispositions to certain diseases such as atherosclerosis, breast cancer, and diabetes, and that people with these predisposing genes will be short-lived on average. We probably all know these sorts of short-lived, heart-attack–prone families, just as we know long-lived families. But aging is more general than susceptibility to particular diseases. It is a general deterioration. Is there any evidence that certain genes or combinations of genes pace our overall aging rate?

The short answer is that we don't yet know. Longevity does seem to be inherited, to a certain extent—just as folk wisdom suggests. But differences among families could reflect only susceptibility to specific disease rather than aging itself. What we would like to know is whether characteristics such as athletic ability and visual acuity also decline more slowly in long-lived families.

One of the first studies of the inheritance of longevity was conducted by Alexander Graham Bell, who in addition to inventing the telephone invented the audiometer, a precursor of the phonograph record, and who designed what was for many years the fastest boat in the world. As has happened to a number of scientists who dabble in gerontology, Bell came to the study of aging late in life, when he faced aging up close and personal. His main contribution was to study the birth and death records over two centuries of almost 9,000 descendants of one William Hyde, who died in 1681 in Norwich, Connecticut.[27]

Bell found that the parents who were long-lived tended to have longer-lived children, just as one might expect, but a huge number of Hyde's descendants seemed to be short-lived, surviving only 30 to 45 years, on average. This was not a long life even by seventeenth-century Connecticut standards, so it is possible that Bell was studying only a particular susceptibility to a single early-life disease. However, a later similar study of a Chinese family from the fourteenth to the nineteenth century, and a third study of many Swedish families between 1500 and 1829, also showed that parental longevity affected how long their children could expect to live. The effects were modest but measurable.

Studies of twins also indicate that longevity tends to be inherited, to a certain degree. Maternal twins, having split from a single fertilized egg, are genetically identical. So comparing how closely maternal twins resemble each other in any trait, compared with fraternal twins, who develop from two different eggs and thus share only half their genes, is another convenient way to look for genetic effects. Such a study of several hundred Danish twins born between 1870 and 1880 who survived at least to adulthood found that about 20 to 30 percent of the variability in life span could be attributed to genetic effects.[28] To put this in perspective, height is about 65 percent heritable in the same sense, and IQ studies typically find that genes explain 40 to 80 percent of variability. So the effect of genetics on longevity is not huge, but it is there. The identical Danish twins differed in their age at death by about 14 years, on average, compared with a difference of about 19 years between two randomly chosen Danes. Such a difference may catch the eye of insurance companies, but it will hardly have other folks drawing up their wills or purchasing grave sites prematurely.

It isn't really so surprising that genetics isn't more of a factor. People, even close relatives, live in many different environments, live many different lifestyles, and die of many things, including accidents. Also, environment is pervasive. In mice, for instance, the sex of one's fetal neighbors while inside the mother's uterus affects the rate of reproductive aging. So heredity can hardly be expected to explain too much. Even under highly controlled laboratory conditions, genes seem to account for only about 30 to 40 percent of longevity differences among mice and among tiny, almost microscopic, worms.[29]

These genetic studies describe the simultaneous effects of many genes. What about the possibility that single genes, such as those recently found to double the life span of simple organisms, could have a major effect on the human aging rate?[30] Could genes with such profound effects be discovered in humans?

The answer is pretty definitely no, unless we choose to believe

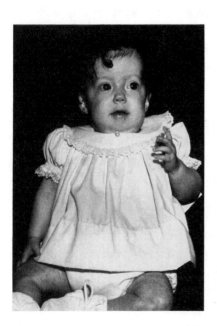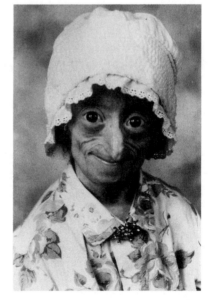

Girl with juvenile progeria, or Hutchinson-Gilford syndrome, at six months and at 16 years of age. She died of heart disease not long after this photograph was taken. (Courtesy of Dr. W. Ted Brown)

45

that some as-yet-undiscovered group of 150-year-olds exists some-where in the world, or unless we are thinking about considerably more modest genetic effects, say 10 to 20 percent. The jury is still out on modest effects. But what about the other direction? Might we learn as much about how we age if we discovered genes that accelerate aging rather than retard it?

We've all seen the photographs in the tabloid press of young children who look like frail old men. At birth, they appear normal, but soon after it becomes evident that they are anything but. They grow very slowly. They become scrawny and bald. Their skin grows thin and spotted; blood vessels are easily seen through it. Heart disease usually kills them by 10 to 15 years of age. This rare disease (1 case per 8 million births), known as progeria or, more specifically, Hutchinson-Gilford syndrome, is caused by a mutation in a single gene. It is popularly thought of as *the* disease of accelerated aging, and suggests that aging itself may be controlled by a single gene.

But is this really aging, or only something that superficially resembles aging? Remember: Aging is a generalized deterioration in which many problems become more common. People with progeria show some, but by no means all, of the signs of aging. For instance, boys with progeria do not develop prostate problems, and progeria sufferers of both sexes are not particularly prone to cancer or cataracts. They also rarely get high blood pressure or stroke or diabetes, and they do not deteriorate mentally or develop Alzheimer's disease. In fact, they may have somewhat higher-than-average intelligence and are fully aware of what is happening to them up to the time they die.

So progeria isn't aging. It only caricatures some outward, and a few inward, signs of it. Other diseases caricature some of aging's other aspects. Werner syndrome, also a very rare mutation in a single gene that has just been identified, is sometimes called adult progeria. People with Werner begin to show signs after puberty. Their hair grays prematurely; their muscles wither; they begin showing signs of atherosclerosis; their reproductive organs fail to develop; and they are prone to many types of cancer. But they are not particularly prone to diabetes, high blood pressure, stroke, prostate problems, or Alzheimer's disease.

In actuality, many genetic problems mimic some aspects of accelerated aging. There are at least 18 different mutations leading to premature loss or graying of hair, another 30 or so leading to accelerated cardiovascular aging, and more than 50 hastening the onset of senility. The commonness of these genetic aberrations has been pointed out by George M. Martin, a geneticist and pathologist at the University of Washington, who is a key scientist in the search for the cause of, and cure for, Alzheimer's disease.

Martin is a brilliant polymath and a human being of exceptional warmth, with close friends in seemingly every country in the world. His home is constantly thronged by visiting scientists, former students and colleagues, and family friends. Unlike many scientists, he has never lost sight of the human pain his research hopes ultimately to remedy. He can be quietly incisive or expansively raconteurial with a devilish sense of humor. He has traveled the world for many years, seeking families with inherited Alzheimer's disease, because studying these families may be our best strategy for understanding the disease.

What Martin appreciated that had previously escaped notice was that childhood and adult progeria were only two of many genetic problems that mimic accelerated aging. He estimates that as many as 7,000 of our 100,000 genes may influence some aspects of the aging rate. He also pointed out that the genetic disease that mimics accelerated aging most thoroughly is not one of the progerias, but a much more common problem—Down syndrome.

Down syndrome occurs in about 1 per 700 births overall, but much more frequently among older mothers. It is caused by an extra copy of one small chromosome (number 21) on which lie many genes, and its physical signs are so characteristic they have been noted in fossilized human skulls. People with Down syndrome have mental deficiencies of varying degrees throughout life, and in the past were stigmatized and neglected to the extent that they were thought to be unable to live past the age of 40. With more enlightened treatment today, however, substantial numbers are surviving beyond 40. Some even live into their seventies.

But many aspects of accelerated aging are associated with Down syndrome. These changes range from the superficial—premature graying or loss of hair, for instance—to the profound, such as a rapid increase in the development of leukemia or vascular disease or the premature development of Alzheimer's disease. People with Down syndrome can develop Alzheimer's as early as 20 years of age, and inescapably develop it by age 35 to 40. But again, other aspects of aging—prostate and breast cancer, high blood pressure, skin wrinkling, osteoporosis, cataracts, and so on—are not accelerated in sufferers of Down syndrome.

Study of these genetic defects, which accelerate some aspects of aging, should have something to teach us about some of aging's specific problems, such as the development of atherosclerosis. And modern molecular biology has provided us with powerful tools for investigating how single genes work. So medical progress along these fronts should be rapid. But at least one other type of genetic difference among people is associated with life span. It isn't a rare genetic problem but rather, a common genetic endowment. That genetic difference is sex.

The Weaker Sex?

Women, to be specific, die at lower rates and therefore live longer than men in virtually every country and culture in the world. In some places, such as Russia, women outlive men by nearly 10 years. The greater longevity of females is seen in places where life is harsh and life expectancy is 40 years, and in the most highly technological of countries where people can expect to live 80 years. The differences between the sexes are smaller in times and in places characterized by poorly developed medical care, but they are still there.

Even among the violent Yanomamö, a girl, once she makes it through her first year of life, can expect to live nearly a year longer than a boy (19.1 versus 19.9 years). Of more than 170 international entities (countries, principalities, islands, etc.) that report demographic information to the United Nations, only six (Nepal, Maldives, India, Iran, Bangladesh, and Bhutan) report greater life expectancy for men than for women, and the difference in these cases is small.[31]

So strikingly unusual is this pattern when found that one immediately suspects that women are treated differently in those cultures than in others. In some cases, it is obvious that the sexes are valued differently. In India, for instance, young girls suffer from malnutrition at four times the rate of boys, but are taken to hospitals only 2 percent as frequently.[32] And even in these countries with greater male life expectancy at birth, women can expect to outlive their male counterparts if they manage to survive childhood and the childbearing years. Only the Maldives, an Islamic republic consisting of about 2,000 small coral islets off the southern tip of India, still reports a marginally greater male life expectancy by the time people reach age 45.

As pointed out previously, women live longer not because they age more slowly. Women simply die at lower rates throughout life—from conception to cremation, if you will. Prematurely born girls even respond better to medical treatment and are more likely to survive than prematurely born baby boys. Because men die at higher rates throughout life, there are three to four female centenarians for every man lucky enough to make it that long. Women, then, are without doubt the biologically stronger sex. The question is why?

People generally dislike and distrust biological explanations of any sexual differences, whether they be in mathematical accomplishment, musical talent, or a knack for catching a baseball. Culture is invariably the prime suspect. But what cultural conditions could similarly affect women in the United States in 1850 and women in Ethiopia, Sweden, and Paraguay in the present so that they live longer? Maybe there are a few causes of death—accidents or murder or heart attacks—to which males are particularly prone, and that could explain their shorter lives. Maybe the explanation is purely behavioral. Maybe men take more risks, in the bush or boardroom, and are done in by excessive stress if they don't die accidentally.

For some things, male behavior *is* no doubt the culprit. Historically, more men than women have smoked cigarettes, and men have typically smoked more heavily. Not surprisingly, then, men die of lung cancer at more than twice the rate that women do. But the gender gap in smoking has been closing in recent years, and the lung cancer deaths among women are rising (in the United States, lung cancer is now the number-one cause of cancer death among women, having passed breast cancer in 1991). Without question, men are also more prone to violence and accidents, particularly in the time of testosterone dementia. But although men may specialize in violent death, succumbing to murder and suicide at more than three times the rate of women, they typically die at higher rates of just about every cause—cancer of almost all types, stroke, heart disease, a host of causes in infancy, even pneumonia and flu. Also it's not easy to imagine how male behavior could reduce the survival prospects of a prematurely born baby boy in a modern hospital.

So it's difficult to escape the conclusion that women are simply better designed biologically for survival. We know that women seem to handle dietary fat better than men. They have lower cholesterol on average, higher levels of the beneficial high-density lipoprotein (HDL) cholesterol, and lower levels of more harmful low-density lipoprotein (LDL) cholesterol. We also know that estrogen (which like testosterone requires cholesterol for its formation) is somehow involved in this state of affairs, as shown by the fact that after menopause, when estrogen virtually disappears from the body, women's cholesterol counts look more like men's. Also, women taking postmenopausal estrogen replacement continue to have healthier cholesterol counts than men on average. So being genetically endowed with lots of estrogen and not much testosterone might somehow explain why women are less prone to heart disease than men, but it doesn't seem to explain less cancer or a greater resistance to death from pneumonia.

Actually, it might. By several measures, women have more effective immune systems than men, at least when they are not pregnant. Pregnancy suppresses the immune system. It is possible that a more vigilant immune system more reliably devours precancerous cells before they become a problem. That could explain the overall difference in the cancer rate between the sexes, and maybe even the survival differences in the face of pneumonia and the flu.

And yet, female mice also have more vigilant immune systems than do male mice. So do female guinea pigs. Mice and guinea pigs, in fact, were the animals in which the superiority of females' immune systems was first described. Yet male mice live just as long as female mice, and guinea boars, if that is the right term, live substantially longer than guinea sows.[33]

At this point, we don't really understand why women live longer than men, although it's hard to discount the immune system's involvement. There is something of a down side to having too vigilant an immune system, though. Like a nervous sentry, it can mistake friends for enemies and shoot the wrong people. So women are more frequently attacked by their own immune systems than are men, resulting in diseases such as arthritis, asthma, and lupus. Moreover, even though they live longer than men, women are more frequently sick and bedridden. They visit doctors and hospitals more often. Men may be more fragile, but they are stoically and stubbornly fragile.

So researchers have the genetic and hormonal tools to start investigating a number of specific diseases associated with human aging. But specific diseases do not equal aging itself. Addressing the general issue of what causes, and paces, the overall aging rate will not be so easy as determining the cause of any specific disease. Future research on aging will require an appreciation not just of what goes wrong during aging but of why it goes wrong in the first place. The next few chapters take up that issue.

5

Why Does Aging Happen?

In almost any other important biological field than that of senescence, it is possible to present the main theories historically, and to show a steady progression from a large number of speculative, to one or two highly probable, main hypotheses. In the case of senescence this cannot be profitably done.

ALEX COMFORT (1956)

Before he discovered the joy (and profitability) of sex, Alex Comfort was the best-known and most astute student of aging in the world. Comfort was an intellectual powerhouse. In a burst of compensatory energy, he published a book per year for a decade after blowing four fingers off one hand in a teenage accident. Although trained as an obstetrician, he wrote novels, poems, and essays in addition to his scientific investigations. Over the years, he compiled information on aging in dogs, horses, snails, birds, and fruit flies, looked at how diet and the ability to regenerate body parts affected the aging of guppies, examined the effects of antioxidants on longevity, and synthesized an enormous amount of research by others to see whether he could make sense from a spotty and fragmented literature. He also founded, and was the first editor of, *Experimental Gerontology*, one of the most prestigious scientific journals in the field of gerontology. And until Caleb Finch's magisterial 1990 tome, *Longevity, Senescence, and the Genome*, Alex Comfort could literally be said to have written the book on aging.[34]

Comfort was right about one thing in the quote above. The number and range of aging theories indeed rather stupefies the imagination. And the list of theories is growing as researchers discover more and more subtle changes taking place in our bodies as we age. For each new bodily change discovered, some scientist envisions scientific immortality (or more likely envisions the humble acceptance of a well-deserved Nobel Prize). So instead of merely being satisfied with having discovered one more *sign* of aging, the scientist claims to have uncovered the *cause* of aging.

How out of control is this proliferation of theories? Not long ago, Zhores Medvedev listed and categorized more than 300 aging theories.[35] But does a huge number of theories really indicate a lack of progress in understanding a phenomenon? Maybe, maybe not. The answer depends on the kinds of theories we are speaking of, and whether they all are operating at the same level of explanation. Let me explain what I mean.

If you ask a physician or medical researcher why we age, the answer will depend a great deal on that person's area of expertise. A neurologist might say that aging occurs because of accumulated damage to neurons—brain cells that are never renewed after birth. On the other hand, a cardiologist might attribute aging to damage in the heart and arteries that gradually reduces blood flow to certain vital areas. A cell biologist is likely to point out how free oxygen radicals, highly damaging molecules produced by normal metabolic processes, injure crucial components in each of our cells until those cells can no longer function properly. And so on. Because there are so many different medical specialties, it isn't difficult to see how one could eventually end up with hundreds of theories. What is more, to a limited extent, many of these ideas may be right at the same time.

However, if you ask a biologist—particularly an evolutionary biologist—the same question, you will find that getting answers isn't so easy. She is likely to start shooting questions right back at you. "What is your definition of aging? Are we speaking on a molecular, cellular, or organismic level? Is that a proximate or ultimate question?"

Evolutionary biologists are not simply obtuse—or, rather, they may be obtuse, just not *simply* obtuse. What they have realized is that not all explanations are equivalent in what they seek to explain. Some explanations are about *how* a phenomenon occurs—that is, by what mechanism it operates. Others are about *why* it occurs. Yet in common language we use the same word, *why*, to address both issues.

To use an automobile analogy, the answer you might expect to a question about why your car moves will depend on whether you ask your local garage mechanic or a physicist. The mechanic will, not surprisingly, give you a mechanistic answer, detailing how electrical energy is transferred from the ignition switch to the solenoid and starter, which turns the engine over. Then, if the car has sufficient fuel, the engine will start, and when the engine is connected to the wheels by engaging the transmission, the car will move. And so on. A physicist, on the other hand, is likely to tell you how the combustion of gasoline in air yields kinetic energy as a consequence of the breaking of chemical bonds, and that this energy can be harnessed by various mechanical devices to do work such as moving an automobile. Both types of answers are correct, they simply apply to different levels of explanation.

This distinction is important in that a mechanic may be able to use the first type of explanation to repair your car, but if he doesn't have some appreciation of what the physicist is saying, then he probably won't be able to figure out any sort of novel mechanical problem. I know this because I worked during the summers of my college years as a truck driver. Wag, the mechanic who was usually sent out to fetch me and to try to fix the engine I had found a novel way to ruin, would softly chant a physics mantra to himself as he tinkered in the bowels of the engine: "We need fuel, fire, and air . . . fuel, fire, and air." A real physicist, on the other hand, may be able to give you chapter and verse on the energetics of vehicular motion, but may not be able to find and fix a spark plug on a bet.

Evolutionary biologists often call the mechanic's type of answer a proximate or mechanistic answer; the physicist's type of answer is called a causal or ultimate answer. My contention is that although there may be more than 300 theories of aging, they are virtually all mechanistic theories, many of which may be correct simultaneously. On the other hand, like the mechanic without an appreciation of simple physics, biomedical researchers are not likely to understand the fundamental processes of aging without an appreciation of the causal answer—why aging occurs in the first place.

Fortunately, understanding why we age is much simpler than understanding how we age, because there are only three causal theories to assess, and two of these have been conclusively disproved. Alas, there is little or no appreciation in much of the medical community that the correct causal theory of aging has now been identified, and this lack of knowledge is no doubt inhibiting progress in answering more mechanistic problems of aging, such as how we might go about slowing its progress.

These three causal theories of aging I will call the good-of-the-species theory, the rate-of-living theory, and the evolutionary aging theory. All the theories have a superficial cogency, which is why they all live on in some recesses of the medical community. But as we will see, two of the theories are clearly false.

Does Anything Evolve Because It Is "Good for the Species"?

The earliest ideas about aging all assumed that any phenomenon so ubiquitous must be somehow advantageous. Because growing more and more feeble with time has no apparent advantage for the individuals to whom it is happening, the logical assumption was that aging must somehow benefit an entity besides the individual. The logical entity was the species.

The logic of this argument is as follows. It is beneficial to a species to be able to adapt to its environment. But environments are continually changing, mainly because the environment consists of other organisms under continual evolutionary pressure to become better adapted. Predators will always benefit from better running ability, as will their prey. Parasites that are more capable of evading the immune system will be at an advantage, and so will animals that develop more effective immunity. In order to keep up with evolutionary changes in other organisms, new mutations and novel combinations of genes must somehow arise. It is only when one generation dies out and is replaced by a new generation that new combinations of genes can arise, allowing increasingly refined levels of adaptation as these new combinations are sorted out by natural selection. If aging did not occur, the argument runs, the necessary turnover of generations would not occur. Evolution, in other words, would slide to a screeching stop without aging and death. Species that failed to evolve would be vulnerable to extinction and would be replaced by other species that were able to evolve. Aging, then, is a phenomenon that evolved because species are better off if they are adaptable.

Virtually *any* theory that posits an advantage to aging will fall into the good-of-the-species category. Some evidence supporting good-of-the-species theory is pretty convincing at first glance. For instance, how else might one explain the sudden, predictable, inevitable death of animals such as Pacific salmon, which are born in freshwater streams, migrate to the ocean as juveniles, and spend a few years there before fighting their way back upstream to the creek of their birth to spawn once and then die within a few weeks?

This mode of life—reproducing in a single, explosive suicidal burst—is called semelparity, and is more common than is generally supposed. In addition to Pacific salmon (Atlantic salmon can breed repeatedly), a large number of plants (which we call "annuals") are semelparous, as are lampreys, certain octopuses, and a number of small marsupial mammals. In the marsupials, which are mouse-like in appearance and behavior but carry their pups in a pouch, all males in a population die of an abrupt immune-system collapse within several weeks after the end of the mating season.

How might semelparity have evolved repeatedly if there were no advantage to the species? More generally, might the same rapid process of deterioration that occurs in semelparous animals occur in all animals at more gradual rates, taking more subtle guises? Could there be something akin to a suicide switch that turns on at some point in life and slowly but inexorably leads to the death of its bearers? This notion has been called programmed aging, because it assumes that the events leading from conception to death follow a genetic program, in which, as in a computer program, one predetermined step inescapably follows another. Thus, aging and death may be nothing more than appropriately timed life events, such as birth, puberty, hair loss, or a fondness for soft chairs.

Before there is any opportunity for confusion, let me say that the good-of-the-species theory is assuredly false and has no proponents among modern evolutionary biologists. Alas, it still has its proponents in the medical community, where evolutionary biology is seldom understood, and even more seldom understood in its modern form. Why medical schools require no literacy in evolutionary biology, even though evolution is *the* grand unifying idea of biology and virtually all medical training is biological in nature, is a question best addressed to medical educators, I suppose. But evolutionary biology does have practical consequences—it might, for instance, have prevented the absurdly futile attempt a few years ago to transplant a baboon's heart into a human infant by suggesting the use of the heart of a much more closely related chimpanzee.

Getting back to the main point: Why is the good-of-the-species theory faulty? There are two reasons. First, its logic confuses death with aging. It assumes that nonaging individuals live forever, but, as mentioned earlier, failing to age is not equivalent to immortality. It is only the persistence of pristine youthful health and vigor for as long as an individual lives. Even nonaging inanimate objects such as test tubes and glass tumblers eventually "die" from accidents. If time doesn't get you, chance will. So evolutionary change will not be halted by failing to age, although it may be slowed.

Second, the theory contains a well-known evolutionary fallacy that is pervasive among the lay public, including the medical community. This fallacy assumes that the traits we find in nature will be the ones that evolved because they were beneficial *to the group*— that is, *to the species or population* exhibiting them. I hasten to say that group selection wasn't always recognized as a fallacy. It was shown to be so during an intellectual battle during the 1960s that followed the publication of a book by V. C. Wynne-Edwards called *Animal Dispersion in Relation to Social Behaviour*. Wynne-Edwards speculated that the rather ridiculous strutting, posing, and chest puffing that we often see animals performing in defense of their territories had evolved as a means by which animals could monitor their own population density. When the density grew to a point that threatened the food supply, individuals would voluntarily restrain themselves from reproducing so that the population would not grow too fast, overeat its food supply, and starve en masse. Individuals that were voluntarily celibate, therefore, were performing an altruistic act to benefit the group—that is, the population or species.

Why was Wynne-Edwards wrong? In order to understand the fallacy of this group selection and recognize it when it crops up in unexpected disguises, we need to quickly review the manner in which evolution by natural selection actually works.

The idea of biological evolution had been in the air for centuries before Charles Darwin's best-selling book, *On the Origin of Species*, appeared in 1859. Darwin didn't invent the idea of evolution, he uncovered a mechanism—natural selection—by which it might plausibly operate. Plausible mechanisms are important in science because professional scientists are skeptics by temperament and training. Unless you can convince them that there is a plausible means by which a phenomenon can occur, they will remain obdurately unconvinced. That is why, as Stephen Jay Gould has pointed out, any dunce can examine a map and see how the humped east coast of South America fits snugly into the concave west coast of Africa, but the idea that the continents were once connected like the pieces of a jigsaw puzzle was not taken seriously until the 1960s. That is when ocean-floor spreading, a mechanism by which continents could move, was finally discovered.

Darwin's chief insight, honestly earned from a lifetime's obsessive and detailed observation of nature, was that individuals are unique. In other words, there is no such thing as *the* rat, *the* dog, *the* human. There are only rats and dogs and people. Individuals within a species have certain similarities, to be sure, but they also exhibit substantial differences. Combining this insight with the Malthusian observation that no plant or animal population can grow unrestrained for long without outstripping its resources, Darwin realized that among all the variable progeny produced within a generation, only those that were best designed for survival and reproduction were likely to form the parents of the next generation. In this way, new beneficial traits, as they arose, would progress from initial rarity to common-ness and finally ubiquitousness within an interbreeding population.

The preceding scenario assumes that the advantageous traits in question are genetic traits—that is, that they are passed from one generation to the next by the replication and transmittal of parents' DNA to their offspring. Genes, which are lengths of DNA, are what is transmitted. And talking about the fate of traits in the context of natural selection is really the same as talking about the fate of genes. Traits (or genes) will generally spread through a population when they are more advantageous than the prevailing trait. New traits arise from genetic mutations, alterations in the DNA, most of which are damaging and soon disappear because their carriers reproduce less successfully than do those without the mutation. But a rare muta-tion that improves the odds of successful reproduction for its carriers will be passed on to the next generation at a higher rate than the prevailing genetic trait is passed on, and as generations pass it will soon become the new prevailing trait.

Most traits that are beneficial to a species will also be beneficial to the individuals that bear them, which is why this issue never came up directly during the first 100 years of Darwinian thought. But not always. Sometimes a conflict will arise between what is beneficial to individuals and what is useful to populations or species. It is these traits that we need to analyze. Which interest will prevail—that of the individual or the group?

Here is how we can think about it. Imagine that there is an optimal population size for a species. At this size, virtually all avail-

able resources such as food and water are used up at exactly the rate
at which they appear, and individuals share these resources equitably, so that there is just enough for everyone—a perfect com-
munitarian paradise. From the viewpoint of the population and its
prospects for persistence, this is an ideal state of affairs. The large
size of the population keeps the group buffered from catastrophic
events, in the sense that the more individuals there are, the more
likely it is that at least a few will survive an environmental catastrophe. If some survive, the population can always regrow to its ideal
size. The only way to perpetuate this state of affairs, however, is to
follow a reproductive strategy of zero population growth when the
population is at its optimum size. That is, every pair of mates must
produce two and only two offspring so that the population will remain at its ideal level, and life in Happy Valley will continue.

From the perspective of an individual, though, the sort of reproductive restraint necessary to maintain this population is anything
but evolutionarily advantageous. If your own particular genes are to
persist—or, better, spread—you should produce as many healthy
progeny as possible as quickly as possible. Therefore, it is advantageous for individuals to produce as many offspring as they can successfully raise, regardless of the current size of the population.

In this sort of conflict between traits that benefit individuals
versus those that benefit populations or species, the key question in
terms of evolution is which traits will ultimately survive? A simple
thought experiment reveals the answer.

Let's imagine that all individuals in our imaginary population have
genes dictating that, in the absence of a catastrophe, each pair of
mates will consume only the resources necessary to produce two
offspring in their lifetime. The population will be stable, and resources
will be equitably shared among parents. Now imagine that a new
mutation occurs in the "reproductive restraint" gene, one that insists
that its carriers produce four offspring, provided that they can ac-
quire sufficient resources by force or guile. This new mutation will
undoubtedly spread, gradually replacing the previous gene, because
its carriers out-produce their more altruistic neighbors. Like thieves
in a nunnery, the carriers of the new "selfish" genes will have little
trouble getting more than their fair share. The fact that these "self-

ish" genes may lead to the food supply's ultimately being overeaten is no barrier to their competitive superiority.[36]

An interesting medical instance of the conflict between groups and individuals, one that follows exactly the same evolutionary logic, is the conflict between the "normal," well-behaved, cooperative cells of which our bodies are generally composed and renegade cells capable of nothing but reproduction—that is, cancer cells.

Think of your body as a happy, enormously large population of different sorts of specialist cells. Liver cells, for instance, are specialized for certain chores, such as removing harmful substances from your blood. Blood cells carry oxygen or provide immunity from infection. Muscle cells contract when commanded to do so by appropriate nerve cells. Some types of cells also reproduce; that is, they divide to produce new cells, but they do so as needed. As skin cells are worn off, underlying cells will divide to replace them, or, if you are wounded, cells will divide until the wound is repaired. But for the most part, cells just perform their specialized tasks. As long as this exquisitely coordinated cooperation continues, you will remain alive and healthy.

However, cancer is the result of new mutations in your body's cells—mutations that cause liver or blood or other cells to cease their normal functioning and just keep growing and dividing. Just as the "selfish" genes that are causing increased reproduction spread and become increasingly common, so these new highly reproductive cancer cells will spread. Of course, at the group level these mutations are anything but advantageous. Once the cancer-cell populations have grown sufficiently large, we (and those cancer cells) die. Although these cells have reached an evolutionary dead end, their short-term reproductive advantage is unassailable. Mutations of this sort are always advantageous at the level of the individual cell, and this explains why cancer is probably the most common disease in higher animals and why it is a particular problem of aging, as we shall see later.

In an analogous case that considers aging per se, imagine a population in which adults rapidly age and die and are replaced by their young, because genetic turnover is good for the long-term adaptability of the population, not because environmental conditions necessitate death. Then a new, mutant longevity gene arises that does nothing but delay aging. The carrier and the carrier's offspring, because they are longer-lived than fellow members of their species, will ultimately reproduce more copiously. Soon, longevity genes will be ubiquitous in the population.

As straightforward as this logic appears, even the most able medical gerontologists find it insidiously easy to fall into the trap of assuming that some patterns of aging exist because of a group advantage. For instance, one notion put forward in the medical literature about why salmon die after their first spawning is that by dying and then decomposing in the stream in which they reproduce, salmon are enriching the stream for the young fry that will soon emerge from the eggs and head downstream toward the ocean.

Why does this logic not compute in terms of evolution? Well, it could, in principle, but only if one assumed that the nutrients released by the decomposition of the parent benefited only its own offspring. Why is this? If the parents' death and decomposition were benefiting other fry in the stream, their death would be enhancing the reproduction of other, competing parents. A mutant individual that lived on, accepting the altruistic assistance by death of others without reciprocating, would ultimately leave more descendants, and soon the selfish, new longevity gene would become common.

If the preceding logic of the inherent success of "selfish" genes is sound, how does one explain the indisputable fact that many animals and plants do reproduce once and immediately die? Is there a possible scenario in which this pattern of sudden aging and death could benefit the individual? Indeed there is. But before elaborating that logic, there is one other fallacy about aging that is widely accepted in the medical community and that needs a bit of explaining.

The Myth of Aging as Limited Cell Division

Several years ago, an Australian entrepeneur had what he considered to be a brilliant notion. He would buy a large painting by Picasso, chop it into one-inch squares, and then sell to an art-starved but penurious public hundreds of original "Picassos," profiting handsomely in doing so. Needless to say, the art world was scandalized. Massive negative publicity soon made his idea untenable anyway. No one but the crassest and thickest-skinned of philistines would have risked the public opprobrium attendant on purchasing *this* sort of Picasso.

But why was the art world so scandalized? Our entrepeneur was doing nothing dishonest. These squares were indeed original Picassos. Or were they? I think that members of the art world objected so vociferously because (among more complicated sociological reasons) they felt that a one-inch painted piece of a larger canvas could not possibly be a Picasso, no matter who originally painted it. Reducing the painting from its integral whole into tiny components would have destroyed its essence. The same may not be true of all paintings. I think of Jackson Pollock's, for instance. But the point is that for anything to have meaning, it must be observed and analyzed at the proper scale.

In modern medical research, the scale of investigatory choice is the cell or an even smaller unit, the cellular component. This is understandable. Compared with whole bodies, cells are relatively simple and therefore relatively easy to investigate. We also have exquisitely powerful molecular tools for probing inside cells, and these tools have largely driven medical research for the past two decades.

But the preferred scale is not always the most appropriate. Sometimes what the philosopher Daniel Dennett calls "greedy reductionism" loses the essence of the phenomenon under study. Studying aging at the level of the cell does exactly this. It may seem obvious that deterioration of complex abilities such as our sprint speed or visual acuity or any of the general debilitations of aging would require changes in many types of cells in various parts of the body. Actual cell death may even be involved in some of these aging changes, as may be the disruption of specific cell functions or perhaps a reduced ability of cells to begin multiplying when needed. However, in 1961 a young cell biologist, Leonard Hayflick, along with a collaborator, Paul Moorhead, made some observations that led a generation of biologists to imagine that they could indeed study aging itself by examining the growth of cells in laboratory dishes. They were wrong, it turns out. What they have been studying these many years is clearly *relevant* to aging and of major medical interest in its own right. It is just as clearly *not* aging itself.[37]

Prior to 1961, the prevailing orthodoxy for half a century had been that even though animals invariably die, the cells of which they are composed could be propagated forever like plant cuttings under the right laboratory conditions. This orthodoxy was largely the work of Alexis Carrel, a Nobel Prize-winning pioneer of cell propagation techniques. A strident, dogmatic, and impatient man convinced of the fundamental incompetence of most other people, Carrel prided himself on having kept cells from the heart of a single chick alive and dividing for decades. Indeed, in the hands of an assistant, one of the cell cultures even outlived Carrel. When other scientists sporadically reported the failure of their cells to continue growing, Carrel claimed that it was *their* errors and incompetence that were responsible. It is now clear that the errors and incompetence were Carrel's own, although a more charitable interpretation is that in his case it was the laboratory assistants who were incompetent or that they spiked the dishes occasionally with fresh cells because they were just too afraid to tell him that the cultures had died.

But Hayflick was a confident young man, not so easily intimidated by Carrel's reputation or by the weight of several decades of entrenched orthodoxy. An exceedingly careful experimenter, he used cells from human fetuses to discover that *normal* cells, as distinguished from cancerous or "transformed" or "immortal" cells, could not in fact be propagated forever. They divide a certain number of times and then forever cease. This limited potential for division, now known as the Hayflick Limit, he interpreted as aging at the cellular level. As in aging generally, a progressive deterioration in the complex machinery of cell division was ultimately assumed to be responsible for halting cell division. Cells sitting inertly in a dish were analogous to dead organisms. Indeed, soon after cell division stopped, the cells began dying. So tracking cells from their initial vigorous state through to their final divisions was the same as tracking a human or animal from early life to death. Apparently supporting this interpretation, Hayflick further observed that cells taken from adults did not divide as many times as did fetal cells, and that cells from relatively long-lived species of animals divided more times than did those from short-lived species. This theory made perfect sense at the time and was a godsend. Here was the whole problem and paradox of aging encapsulated at a level we could work with.

Before continuing, let me emphasize that this theory of cellular aging is a mechanistic theory, not a causal one. Therefore, it doesn't directly compete with any of the causal theories.

So why do I say that the Hayflick phenomenon is not equivalent, but only relevant, to aging? That even though major medical textbooks often cite the Hayflick Limit as the fundamental essence of aging, in reality that essence has been lost in reducing the problem to this scale?

First, and most obviously, there is no evidence that the generalized bodily deterioration that we call aging results from the inability of previously dividing cells to divide any longer. Take some fairly general aging change, such as the 1 percent loss of muscle strength per year that we suffer as we grow older.[38] Muscles grow weaker as we age not because muscle cells no longer divide. They don't divide after being fully formed during development, anyway. (Muscle size increases with resistance exercise because individual cells increase in size, not because new cells are produced.) No, senile muscle weakness is caused by the death of some muscle cells and a weakened ability to contract in the cells that remain. Also, people don't usually die because their cells have stopped dividing; if anything, they die (due to cancer) when their cells continue to divide inappropriately.

This is not to say that people *can't* die or develop diseases because of limited cell division. Some recent research, in fact, suggests that limited cell division may inhibit healing in arterial regions that have been damaged by being bombarded with blood.[39] Also, our immune system needs rapid, plentiful cell division to operate effectively, so limited division potential may be involved in the reduced immunity of the elderly. But is it involved in all aging changes—in the muscles, eyes, and brain?

In addition, we have recently learned a number of new things about these no-longer-dividing cells. They don't necessarily die, for instance, once cell division has stopped. Although the earlier research observed that cells began dying soon after they had ceased dividing, better cell-culture conditions have shown us that this needn't be so. Cultured cells can live at least several years beyond the end of division. Even then, they usually die because the cultures get contaminated rather than because the cells have become terminally damaged. Also, we now know that key functional changes in these cells, particularly the active turning on of genes that suppress cell division, appear to be part of normal cell function, not specifically the result of accumulated damage. Damage may hasten the ceasing of division, but it doesn't appear to be essential for it. These cell-replication–suppressing genes are turned on even in some fetal cells—that is, before aging has begun—not just in the cells of adults or the elderly. Besides, if the loss of cell-replication ability were indeed equivalent to aging, why would evolution have designed an active mechanism to initiate an ultimately destructive event such as aging? Except for erroneous good-of-the-species arguments, there seems to be no rationale that makes even rough sense.

But if the loss of replicative ability in cells is not aging itself, what is it? And why do we observe that cells from short-lived animals and adults seem to be closer to the end of their replicative lives than those from long-lived animals or fetuses?

The answers to these questions can be understood in evolutionary terms. In order to thrive in a Darwinian sense, animals need to do at least two things successfully. They need to develop from a single cell, the fertilized egg, into a trillion-celled normal adult body form; and they need to avoid premature death from disease. Limited replicative ability is the key to both of these activities.

Imagine trying to create the complex form of the human body through the repeated division of what was originally a single spherical cell. Long appendages such as arms and fingers must eventually appear, as well as bag-like lungs, stomachs, and bladders, not to mention the lunar landscape of convolutions of the brain. The only way to get from there (the egg) to here (the body) is to have some cells divide an astronomical number of times (but not infinitely), others stop dividing at some relatively early stage, and still others die prematurely. Limited cell division is one of the requirements for proper development. So it's not surprising that fetal genes are activated to stop cell division in some parts of the body. On the other hand, once a fully formed adult has been created, a major danger to continuing survival is cancer—uncontrolled and inappropriate cell division. Wouldn't it make sense for evolution to design a fail-safe system, limited cell-replication ability, to help keep adult cells in check?

Considering the Hayflick Limit as a necessary part of growth and development and as a protective device against cancer helps explain the patterns that seemed so convincingly to link limited cell division to aging in the first place. Adults, having used up much of their cell-replication potential in growing into adulthood, would be expected to have less capacity for division. So cells from adults divide fewer times in a laboratory dish than do those from fetuses. Similarly long-lived animal species, which are generally large, require more divisions before shutting down than do short-lived species, which are generally small, because large long-lived animals such as elephants consist of many more cells than do small short-lived animals such as mice, even though large and small animals alike begin as a single egg.

The Hayflick Limit is relevant to aging, even though it is not the thing itself, because of its role in preventing cancer. One of the most general and inescapable problems of aging in complex animals, from frogs and lizards to mice to humans, is cancer. Understanding how nature limits cell division is helping us understand how cancer can develop, and it will ultimately help us prevent its development.

In fact, we are just now closing in on one type of such natural limit—the way that cells keep track of the number of times they have divided so that they will divide enough times to do their job but still be protected against the limitless cell division of cancer. The keys appear to be what are called telomeres and a recently discovered enzyme called telomerase. Telomeres occupy the ends of all our chromosomes. They are short sequences of so-called junk DNA repeated hundreds to thousands of times, and were previously thought to have no function except to keep chromosomes from becoming stuck together end to end. Each time a cell divides, all its chromosomes must replicate so that the two daughter cells each have all the same genes as the parent cell. However, because of the mechanics of DNA replication, one strand cannot be replicated quite to its very end. Thus, chromosomes in reality become slightly shorter with each cell division, as if small bites were being taken from the ends. Having telomeres turns out to be useful, because having fewer of these long, functionless sequences of DNA will not affect the working of the cell. Interestingly, it now appears that the length of the telomere is critical to stopping runaway cell division, because when the telomere shortens by a critical amount, it activates genes specialized for shutting down cell division. So the number of divisions any cell undergoes seems to be limited by the original length of its telomere.

However, this nicely designed fail-safe system can sometimes fail, because not all cells can afford limited division potential. For instance, reproductive cells (sperm and eggs) must be able to divide essentially forever, because our parents' reproductive cells, besides dividing countless times after fusing to form us, also will ultimately become *our* reproductive cells. Similarly, *our* reproductive cells, after countless divisions, become our sons' and daughters' reproductive cells. If reproductive cells didn't have some special mechanism to avoid telomere shortening, then each generation would have shorter telomeres than the previous one, until reproduction stopped because no new reproductive cells could be manufactured. Reproductive cells solve this problem by manufacturing an enzyme called telomerase.

Telomerase adds back the shortened ends of telomeres after each cell division, hence allowing limitless cell proliferation. Cells that do not need infinite cell division normally have their telomerase-production genes turned off; reproductive cells have them turned on. This is our Achilles' heel, because damage to the genes that suppress telomerase production in normal cells takes the brakes off. The cell now becomes a potential runaway train.

The "up" side of this news, which makes telomerase so interesting to cancer researchers, is that if we can find a way medically to turn off telomerase, it may allow us to arrest cancer growth using one of our bodies' own products.[40]

Getting back to aging, although some scientists persist in thinking that cellular aging is really aging itself, they are now a beleaguered minority. At a gala celebration of the thirtieth anniversary of the publication of Hayflick and Moorhead's original paper, the gerontologist George M. Martin took an informal poll, asking how many of the scientists present, most of whom had spent a fair fraction of their professional careers investigating the Hayflick Limit, thought that limited cell replication was a normal physiological process and how many thought it degenerative and pathological—that is aging. Only two of those polled still thought that limited cell replication was aging; the great majority believed it to be normal physiology— though physiology of exceptional interest. And at least one of the two holdouts has since admitted to me over dinner and drinks that he has changed his mind.

Having dispensed with the good-of-the-species theory, let us now consider a second causal theory of aging—a theory with a long list of eminent adherents among physicians and researchers. It is by far the most pervasive, plausible, persistent, and intuitively satisfying theory of aging. It is also demonstrably false, an awkward fact that hasn't seem to dull public or professional enthusiasm for it.

6

The Rate of Living

A wren lives three years, a dog three times as long as a
wren, a horse three times as long as a dog, and a man
three times as long as a horse.

AN OLD GERMAN SAYING RECORDED BY JACOB GRIMM

House mice have rapid, short lives; turtles have slow, long ones.
Mice grow to adulthood in less than two months and live only
a couple of years or so even when they are fed the best scientifically
designed diets in hygienic, climate-controlled laboratories. Blanding's
turtles the size of dinner plates, by contrast, take some 15 years to
reach adulthood and can live more than 70 years even in harsh and
hostile natural conditions. Some deep message about the paradox of
aging—that animals, for all their ability to heal wounds and injuries,
inevitably deteriorate over time—must be concealed in the fact that
some species flame out and die young like mice, whereas others, like
turtles, amble through decades of life at a more stately pace.

An idea that has lurked in human intuition—in other words, in
our collective unconscious—for centuries is what we call today the

rate-of-living theory: the idea that aging results literally from the pace at which life is lived.

The pace of life has a nice definitive sound to it, but exactly what does it mean? Does it mean that birds such as swifts that spend virtually their entire lives in rapid flight will be shorter-lived than sedentary sloths? Or that harried executives will be shorter-lived than librarians? Can we expect that the stereotypically serene Polynesian relaxing on a breezy tropical beach will be longer-lived than an overworked hod carrier? In a word, yes. This is exactly what it means to some people.

One way to think about the rate of living is to consider that mice don't simply grow up and grow old faster than turtles; they do everything faster. They move faster, grow faster, breathe faster, digest their food faster, and have a faster heart rate. Why? Chiefly because they burn energy faster and live at a higher body temperature than turtles. Rapid energy use and high body temperature mean that the necessary biochemical events of life proceed more quickly. The rate of energy use by any animal depends partly on its body temperature (the rate of chemical reactions rises two to three times for every $10°C$ increase in heat) and partly on its genetic heritage. Most mammals, for instance, live at about the same $37°C$ (98.6°F) body temperature that we do, but the rate at which different species use energy varies enormously. You can make a quick estimate of how quickly animals burn energy by determining how much they need to eat to keep from starving. Shrews, for instance, have a very high metabolic rate and must eat about their own body weight in food each day. They can starve to death during a long nap. On the other hand, turtles and even some mammals, such as large whales, can go months at a time without eating a thing.

The rate-of-living theory specifies that the rate of energy use—that is, the metabolic rate—and the consequent speed of biochemical activity is what causes and calibrates aging. The not-so-hidden message here is that life itself is inherently destructive and self-limiting. Put another way, it is a theory of biochemical imperfection. Accompanying the biochemical processes of life will be unavoidable collateral damage—perhaps proteins involved in a reaction will be damaged, say by free radicals; perhaps toxic byproducts of a reaction will fail to be degraded; perhaps inert waste products will accumulate, gradually cluttering up our cells until they can no longer function properly. When enough cells are sufficiently damaged, we will die. If animals were cars, we might say that they all start with some amount of fuel. If they use it rapidly, they die early, as mice do. If they use it slowly, they may live as long as turtles.

Turtles do in fact burn energy much more slowly than do mice. Cold-blooded animals—that is, most insects, reptiles, fish, amphibians, and others that must obtain their body heat from the environment rather than generating it internally as do birds and mammals—*do* generally burn energy much more slowly. Why else would people try to farm iguanas?

Iguanas are lizards about the size of small rabbits. Iguana meat, like all unfamiliar white meat, tastes like chicken. The tail is a particular treat. Because they don't squander energy maintaining a high body temperature, iguanas devote more of the energy from the food that they eat to growing—or, as iguana farmers prefer to think of it, to producing meat. On a given amount of lettuce, that is, you can raise many more pounds of iguanas than rabbits. This fact appeals to cost-conscious farmers. On the other hand, people haven't been all that eager to eat something that looks like a small green dragon with a hangover. So in spite of its sound biological sense, iguana farming has not been a raging success so far. But iguanas *do* live longer than rabbits, just as the rate-of-living idea would have predicted.

As much fun as they might be, these comparisons of individual species are really no more than anecdotes. Turtles and iguanas live longer than do mice and rabbits. Big deal. Is there a more systematic and incisive comparison that we can make?

The first scientist to investigate the rate-of-living idea in any rigorous fashion was the German physiologist Max Rubner. Rubner could make people very uncomfortable with his Teutonic bluntness. He was noted for his long silences, punctuated by outbursts of aggressively sarcastic humor. But he was also an obsessively precise investigator of the energy contained in food and the use of that energy by animals. Like obsessives everywhere, he felt that the significance of his obsession was underappreciated by others. Rubner *knew* that energy use explained many aspects of animal life, including maturation and the aging rate, and to demonstrate that this was true, he set about measuring metabolic rates in five species that he could easily get his hands on: horses, cows, dogs, cats, and guinea pigs. What he found was startling and directed to a considerable degree the course of gerontological research for nearly a century.

Although these five species differed a thousandfold in weight (from the one-pound guinea pig to the thousand-pound horse), and fivefold in how long they lived (from the six years of guinea pigs to the 30 years of horses), they each consumed about the same amount of energy per lifetime.

Let's be clear about what this means. It doesn't mean that you can feed a horse the same amount of food that you feed a guinea pig. You would have either a very skinny horse or a massively obese guinea pig. It means that a thousand pounds of horse needs far less food than a thousand pounds of guinea pig. It is necessary in comparing metabolisms of different animals to standardize the measurements by the animals' body weight. There is a good reason for this. The biochemical reactions of life occur inside cells, of course. When we standardize total body metabolism among species by body weight, what we are really doing is standardizing energy use or biochemical activity *per cell*, because as surprising as it may seem, mouse cells, with a few exceptions, are about the same size as human or whale cells. If life is limited by the total metabolism that a cell can support, as the rate-of-living theory claims, then comparing per-cell metabolism ought to reveal this. Rubner found that guinea pigs burned about 260 Calories per gram of body tissue during their lives (1 Calorie = 1,000 calories), horses burned about 170 Calories per gram; and the other species fell somewhere between. Actually, if we calculate the same figures using more recent information on the maximum longevity of horses (now considered to be 50 years), the comparison becomes even more strikingly similar—260 Calories for a guinea pig versus 280 Calories for a horse.

But why should life span be limited by energy consumption? Is there something intrinsically damaging about using energy, and, if so, why can't our bodies repair the damage?

There is good reason to imagine that metabolism can be damaging. After all, metabolism is nothing more than a cold, controlled fire; that is, metabolism is like a fire that uses oxygen to turn organic molecules of wood into carbon dioxide, water vapor, and energy in the form of heat. In the case of our bodies, metabolism uses the oxygen to turn the organic molecules of food we digest into the carbon dioxide, water, and energy we need to continue operating. The main difference between fire and metabolism is that metabolism releases energy more slowly and in a more highly controlled way.

But fires—even controlled fires—can be damaging. Automobile engines, which also use the controlled release of chemical energy to operate, have a finite lifetime as a consequence of the damage and debris created by the fires that burn inside them. So it isn't a huge logical leap to imagine that our internal fires inevitably cause damage, too, and that the damage may actually be what we call aging.

Rubner's work wasn't the only provocative study on how the aging rate and the metabolic rate might be related. It was only the first. It turned out that a wide variety of insects lived longer when their body temperatures were reduced by cooling the rooms in which they lived, and that they were shorter-lived when placed in a warmer room. One study even counted the heartbeats of transparent water fleas swimming slowly in cold water and those swimming rapidly in warmer water. Not surprisingly, the water fleas lived much longer in the colder water. What surprised the experimenters was that the fleas' total number of heartbeats per lifetime was similar regardless of the temperature.[41]

The man who put the rate-of-living theory on the scientific map for good, the man who tested it most thoroughly and gave it a catchy name (something never to be underestimated in evaluating the life span of ideas), was Raymond Pearl. Pearl towered above his peers, literally and figuratively. Unusually tall and broad, he was intellectually imposing, too. He spent most of his professional life at Johns Hopkins University in Baltimore, where he worked on problems in genetics, behavior, fertility, mortality, morphometrics, human diseases, and too many other subjects to name. His laboratory experiments dealt with worms, fruit flies, cattle, corn, cantaloupes, and chickens. He wrote for the lay public on whatever popped into his mind—

eugenics, race, economics, or the effects of alcohol and tobacco on longevity. In fact, in 1938, he published what might have been the first paper analyzing the extent to which smoking reduced life expectancy. In his lifetime, he published 17 books and more than 700 scientific papers, and he founded two scientific journals, which still exist. He also wrote for newspapers and literary journals, and when he died in 1940 at the age of 61, he was eulogized by his closest friend, the journalist H. L. Mencken, who proudly pointed to Pearl's "relentless and effective war upon bogus facts and false assumptions within his own field. He cleared off, in his time, an immense mass of rubbish."[42]

Like Mencken, however, Pearl was dogmatic, intolerant, and overbearing, and he produced quite a bit of his own rubbish. In 1920,

Raymond Pearl, staunch advocate of the rate-of-living theory of aging. Here Pearl is younger than the age (50) after which he thought people should forfeit their right to vote, because they would have grown too foolish. (Courtesy of the Alan Mason Chesney Medical Archives of the Johns Hopkins Medical Institutions)

for instance, he claimed that the United States could never support a population larger than 197 million, and that even that population would not be approached for about 200 years. (In actuality, this population size was surpassed less than 50 years later.) In another rather spectacular mistake, his flawed analysis of Johns Hopkins's autopsy records led him to believe that cancer could be cured by injections of tuberculin, a compound derived from the bacterium that causes tuberculosis. He followed this idea far enough to actually inject several terminally ill cancer patients with tuberculin and was rather pleased with the result, even though the patients still died and subsequent research has shown that there was nothing to the idea.[43] Another idea of Pearl's that may or may not seem to have been a mistake, depending on one's interpretation of modern politics, was to have an upper limit of about 50 on the voting age, because beyond that age he felt that people might be too foolish to vote wisely. How opinionated was he? Even one of his *friends*, the leader of a campaign (subsequently lost) to have Pearl appointed to a prestigious position at Harvard, said of him, "[I]t has been the characteristic of my experience with Pearl that you can't help him. He is so sure that he is right or at least so unwilling to admit that he could be wrong that you can't set him on a better track."[44]

And there is nothing of which Pearl was more certain than that aging was the inevitable offshoot, and could be simply explained by, the rate of energy metabolism. The headline to an article he wrote in the *Baltimore Sun* in 1927 was "Why Lazy People Live the Longest." In fact, he went further and even attributed the greater longevity of women compared with men to the fact that they supposedly performed less physical labor—a statement that from a sedentary academic took a considerable amount of gall.

Pearl had his reasons for holding to this idea, of course. The most obvious was that by this time, a number of scientists had demonstrated that reducing the temperature (within reason) at which cold-blooded animals were living, and by doing so reducing the rate at which they were expending energy, prolonged their lives. Second, in experiments with cantaloupe seeds, Pearl himself found that the ones that grew the fastest, and therefore presumably used energy the most quickly, died the soonest.

Finally, and most impressive to Pearl, was the information he had managed to unearth on deaths of Englishmen and Welshmen in differing occupations. He expected, of course, to find that men in jobs requiring the greatest expenditure of physical energy—say, coal mining and bricklaying—would be the shortest-lived, and that men in jobs requiring the least physical work—say, college professors and accountants—would be the longest-lived. The rest of the occupations, he expected, would fall in between, according to their relative arduousness. He even managed to subtract from the mix accidental deaths and those due to particular occupational hazards, because they were recorded separately.

The only apparent difficulty in what might otherwise have been a definitive study of more than 130 occupations was how to rank occupations by the amount of physical labor they involved. Does a foundry worker, for instance, expend more energy per day than a chimney sweep or blacksmith? What about a bus driver versus a photographer or railway ticket taker? In order to rank the occupations, Pearl sent a questionnaire to eight of his statistically inclined academic friends, asking them to categorize the professions according to the physical labor they required. If I had to choose a group of people guaranteed to know as little as possible about whether bargemen labor more or less than blacksmiths, it would be academics, whose idea of what, say, dockworkers or gamekeepers actually do during the day is matched only by their grasp of, say, the anatomy of Bigfoot. In their final listing, the academics rated lawyers and priests as having the least arduous indoor jobs, and ironworkers and blacksmiths as having the most arduous. With respect to *outdoor* occupations, insurance agents (Outdoor? Things were obviously different back then!) and coachmen were deemed to have the easiest jobs and coal heavers (a delightful-sounding occupation, eh?) and coal miners the hardest. Of course, Pearl, being Pearl, did not agree with all of his friends' placements of these occupations, so he shifted them around as he saw fit.

His results were very convincing—to him. Lawyers and priests from age 55 on indeed died at lower rates than did ironworkers and blacksmiths. The same was true of insurance agents and coachmen versus coal heavers and miners. The differences were particularly stark at ages 75 and above.

But it occurred even to Pearl that differences other than the amount of physical work distinguished lawyers from blacksmiths. Diet no doubt differed among socioeconomic groups at the time, as is true today; the same is likely to have been true of smoking and drinking habits, exposure to disease, and access to medical care, to name a few. Pearl attempted in some fashion to take these other differences into account by considering social class (members of the upper and middle classes versus unskilled workmen and several others) in his analysis. Yet there weren't many upper-class coal heavers or lower-class lawyers available to allow solid comparisons of social class alone. So ultimately, the meaning of Pearl's study was anything but certain. About the only thing these data reveal to a modern eye is that it was indeed possible to work a Welsh coal miner to death.

But Pearl thought otherwise. Despite the slop in his evidence, he was convinced that he had demonstrated that hard work and short life were inevitable companions. I can see how this would be comforting to an academic. However, the implications of his idea are staggering. Humans from primitive or poor cultures expend an enormous amount of energy in the physical labor of their normal lives. Does this mean that they are inevitably condemned to short lives? Pearl assumed that labor per se, not the living conditions of poverty, were responsible for short lives. He even quotes with great satisfaction a book on Chinese culture,[45] in which a physician states that the heart problems of coolie laborers were due to excessive burden bearing. On the other hand, in cultures such as ours, with plenty of wealth and an excess of leisure, some people voluntarily seek to increase their rate of energy expenditure in order to feel and look good. We call this exercise and assume that it is healthful. We will examine the evidence for this assumption later. But if the rate-of-living theory is valid, and increasing energetic expenditure indeed hastens aging, you would expect fitness enthusiasts to pour out of health clubs faster than clowns out of a circus car. Indeed, soon after Rubner's and Pearl's original work, when rate-of-living ideas were virtually unchallenged, several studies found that rats leading sedentary lives in cages lived substantially longer than those given exercise,[46] apparently confirming the rate-of-living theory.

The rate-of-living idea certainly seemed to explain one other observation, too—one that had been made since ancient times: that large animals generally lived more slowly and longer than small animals. Mice grow to adulthood in two months and live two to three years, dogs mature in six to eight months and live 10 to 20 years, and horses reach puberty in about a year and live 30 to 50 years.

As metabolic studies progressed after Rubner's original work, several patterns became clear. First, if you measured the amount of energy animals burned per cell (or per gram) while resting, assuming that this was the minimum amount necessary to sustain life, it turned out that larger animals burned energy more slowly. This relationship was fairly precise. If you compared two animals that differed 16-fold in weight—say, a mouse relative to a rat, or a cat to a chimpanzee—the per pound resting metabolic rate of the larger species will be about one-half that of the smaller species. This relationship holds for mammals from the size of a shrew (a fraction of an ounce) to the size of an elephant.

George Sacher, a statistician and biologist working for the U.S. Atomic Energy Commission at Argonne National Laboratory, was the first to calibrate the details of how body weight, metabolic rate, and longevity were related in a large selection of mammals. Sacher originally became interested in aging research because of the superficial similarity between the effects of low radiation doses and aging. When it became obvious that aging and radiation damage were not similar phenomena, his superiors became disenchanted with his research, but he was hooked on the paradox of aging and persisted with it in spite of official resistance. He eventually became an exceptionally well-liked and respected member of the research community of gerontologists. He even served as president of the Gerontological Society of America at one time.

In the late 1950s, Sacher gathered information on dozens of species of mammals and found that the increase in longevity found with increasing body weight matched almost exactly the decrease in metabolic rate. Because smaller mammals also have faster heartbeats than larger mammals, a folk version of Sacher's finding arose stating that all mammals had about the same number of heartbeats per lifetime. Could it be that the energy rate per cell per lifetime was indeed constant, and that the paradox of aging could be reduced to nothing more than the rate of energy use, as Rubner had suspected many years earlier?

The final, seemingly decisive bit of evidence supporting the rate-of-living theory was provided by a veterinary nutritionist from Cornell University, Clive McCay, who in 1935 quietly published a paper that would alter the course of aging research for decades. Without much fanfare—no press releases or hastily called news conferences—he reported that he had discovered a way to slow aging in laboratory rats. It required no new drugs, no expensive spas or meditational, inspirational, or colonic therapy. All you had to do was force rodents to eat less than they wished to.

McCay had reason not to trumpet his discovery too loudly. The history of aging research had been littered since at least the time of the Greeks with bogus ointments, poultices, and other treatments that supposedly extended youthfulness, retarded aging, and lengthened life. The only trouble was that these claims never held up to careful experimentation.

McCay's experiments weren't motivated by the rate-of-living theory. He had not been in the mainstream of research on aging, and so had no predisposition to think primarily about the relation between energy use and aging. He was more interested in, and impressed by, the relationship between longevity and the rate of development. McCay like others before him had noted that long-lived species reached puberty at later ages than short-lived species, and he therefore wondered whether delaying puberty might also retard the aging rate. Since he also knew that well-fed animals reached puberty quickly, whereas poorly fed animals reached puberty much later, if at all, he hypothesized that reducing animals' food—putting them on a severe diet, in other words—might retard aging.

There even existed a small bit of experimental support for this idea. Partially starving fruit flies or water fleas, for instance, made them develop more slowly and, at the same time, live considerably longer. Even though McCay wasn't thinking along those lines, we should note that this research on hungry insects was perfectly consistent with the rate-of-living theory, because it was also known that putting any cold-blooded animal on a diet reduced its metabolic rate. Some ticks, in fact, are capable of surviving years without eating, because they can drop into a state in which their metabolism is so low that it is barely detectable, more like the metabolic rate of a stone.

If partially starving cold-blooded animals slowed their aging by reducing their metabolic rate, there didn't appear to be much hope for similarly retarding aging in rats or humans. After all, the metabolic rates of cold-blooded animals were highly flexible, determined almost entirely by the environment. You can make such animals live longer by putting them in a refrigerator, where their metabolic rates plummet, and shorten their lives by accelerating their metabolism in warm incubators. Humans—in fact, mammals generally—don't operate this way. We are stuck with that 98.6-degree body temperature, which requires a high metabolic rate to be maintained. Putting humans and most other mammals in a refrigerator *increases* their metabolic rate as their bodies burn fuel faster to keep warm.

But McCay and others who were convinced that aging was caused by the pace of growth rather than the pace of metabolism were not discouraged. Regardless of its effect on the metabolic rate, reducing food intake, they knew, slowed growth and development in adolescent rats, mice, cows, and mammals generally. It is also true for humans. People living in subsistence cultures in Africa and New Guinea grow and mature more slowly than do humans in cultures of superabundant food, such as ours. Due to the low number of calories they eat and the amount of work they typically need to do in order to survive, women in parts of New Guinea do not begin menstruating until 18 or 19 years of age, and don't have children or quit growing until their early twenties.[47] Compare this with girls in the United States and Europe, who now begin menstruating before age 13, often have children in their mid-teens, and stop growing by age 16 or so.

But despite their slow development rate, people living in subsistence cultures are not especially long-lived. They are usually exceptionally short-lived, in fact, and even in McCay's day, putting rodents on severe diets seemed as frequently to shorten their lives as to lengthen them.

McCay's fundamental insight, what set his experiments apart from those of his predecessors, was that he knew there was a difference between restricting the number of calories eaten—that is, dieting—and malnutrition, or starving. Dieting is eating fewer calories but consuming enough crucial nutrients, such as protein, vitamins, and minerals, to sustain normal vital processes such as wound repair, brain activity, and cardiac and immune function. McCay felt that scientists previously had been studying malnutrition. Being a nutritionist himself, he was careful to feed his rats a reduced diet, but one that was rich in vitamins and minerals. Lo and behold, when he did this, his dieting male rats lived about 75 percent longer than did the fully fed animals, on average, and the longest-lived dieting rat lived more than a year longer than its gluttonous counterpart.

McCay found no similar life extension for female rats in his first experiment, but several years later, when he repeated his experiments using a slightly different vitamin mixture, the females did indeed respond by living longer. Since then, these diet studies have been repeated hundreds of times in both rats and mice, and the results have been very consistent. Reducing calories by 30 to 40 percent from the number of calories that these animals would eat if food were unlimited extends life by about 20 to 40 percent. Such a diet also retards the development of tumors and improves an animal's health in virtually every way that can be medically measured.

By the 1970s, when enough replicated experiments confirmed beyond dispute that dieting retarded aging in rodents, George Sacher felt compelled to bring that finding into line with the rate-of-living theory. He assumed that food-restricted rodents were in fact dropping their metabolic rate per cell the way fruit flies do, and that this was what explained their extended life. He even made some back-of-the-envelope calculations from one paper on dieting rats, which reported both how much the animals weighed and how much they ate. From these numbers he concluded that the metabolic rate per cell was reduced by almost exactly the same percentage as life was extended. Therefore, the long-lived, dieting rats expended almost the identical amount of energy per cell in their lifetime as did the normal, fully fed rats. It was a perfect rate-of-living result, which we can now envision by thinking of a huge pile of food containing all the calories our bodies can use in a lifetime. Each individual will live as long as it takes to consume this food. We can consume it all in 40 or 50 years and drop in our gluttonous tracks, or we can, with great torment and restraint, stretch our allotment to last 80 or 100 years. The rate-of-living theory is a perfect Puritan's view of life.

So energy use seemed to be sufficient to explain the differing aging rates of cold-blooded animals living at different temperatures, animals fed lots of food versus those fed little food, and large animals versus small animals. It even seemed to explain certain departures from the general pattern of large, slowly aging animals versus small, rapidly aging ones. Cold-blooded animals in anything but the most torrid environments burn energy much more slowly than birds or mammals, somewhere between one-seventh and one-thirtieth as fast if body weights are the same. And cold-blooded animals are generally pretty long-lived, regardless of body size. Toads and crayfish live longer than dogs, salamanders longer than mice, snakes longer than rabbits, and turtles longer than just about anything. What's more, some mammals with exceptionally slow metabolic rates for their size, such as the primitive, egg-laying spiny anteaters or the slow-moving sloths, live considerably longer than you might expect from their body size alone.

In order to easily compare animal longevities relative to what you would expect from their body size, I developed the idea of the longevity quotient, or LQ. Just as IQ, or intelligence quotient, measures departure from an average mental ability (or at least whatever ability is measured by IQ tests) for a specific age, LQ measures departure in longevity from that expected for a specific body size. An LQ of 2 means that an animal lives twice as long as expected considering its body size alone. By this measure, spiny anteaters live more than three times as long as expected and sloths more than twice as long as expected.

In a scientific coincidence of note, at about the same time that Sacher was doing his calculations of lifetime energy use in various animals, a physician and chemist named Denham Harman began receiving considerable attention for providing what seemed to be a highly plausible *mechanism* for the way that metabolism and the aging rate might be related. Harman advanced the idea that aging was due to cellular damage caused by free radicals, highly reactive molecules that are an inevitable byproduct of normal metabolism.

Most animals, like ourselves, require oxygen for the maintenance of life. It is an elixir without which we cannot burn our fuels effectively. But oxygen is a poison, too. Breathing pure oxygen at normal atmospheric pressure causes chest soreness, coughing, and a sore throat in as little as six hours, and longer exposure than this leads to inevitable, irreversible lung damage. Lungs are not the only tissue that oxygen damages. In the 1940s, when premature babies were generally kept in oxygen-rich incubators, many were blinded by retinal damage caused by oxygen. Oxygen breathed at greater-than-normal atmospheric pressure causes convulsions, and in a laboratory culture dish, human and other animals' cells are quickly killed if grown in high oxygen concentrations. Whole animals, from insects to fish, mice, and rabbits, survive less and less well as the oxygen level in which they live increases.

Oxygen is so damaging for the same reason that it is so useful. It reacts with many other molecules at biological temperatures. If these chemicals happen to be the fats or carbohydrates that provide our fuel, we benefit from their energy-releasing breakdown. If they are molecules that serve some other function inside our cells, then their breakdown is likely to be harmful. Life indeed seems to be inherently destructive and self-limiting.

Oxygen creates free radicals and other damaging molecules. Free radicals are molecules that contain unpaired electrons, which often make these molecules highly reactive. Electrons are most stable when they occur in pairs. If they occur singly, they tend to rip electrons (and atoms) loose from other molecules, damaging them in the process. While this electron theft stabilizes the original free radical, it creates another free radical, and that one may become even more highly reactive than the original, creating another free radical in turn. Through this process, cellular damage occurs and is perpetuated.

Longevity quotients for some selected mammals[48]

Animal	Maximum Zoo Longevity (years)	Longevity Quotient (LQ)
White-eared opossum	4.0	0.3
Asian house shrew	2.5	0.4
Asiatic wild dog	16	0.5
Giraffe	36	1.0
African lion	30	1.1
African elephant	60	1.2
Hippopotamus	54	1.2
Dog	? → 34	1.9
House cat	28	2.0
Southern flying squirrel	17	2.7
African collared fruit bat	23	3.1
Vampire bat	19.5	3.5
Human	90	4.2
Greater horseshoe bat	26	5.0
Little brown bat	32	5.8

Our bodies are designed to manage most normal effects of oxygen. However, several other forms of oxygen, collectively called by biochemists *oxidants*, also have damaging side effects because our bodies do not seem designed to cope with them so well. Normal metabolic processes create these reactive oxygen forms as inevitable byproducts of the burning of oxygen, and these oxygen forms are now thought to be involved in many of the degenerative changes of aging, including atherosclerosis, arthritis, cataracts, and the development of cancer, or carcinogenesis.

If free radicals are in fact solely responsible for aging, and if they are produced in proportion to metabolic rate, then the paradox of aging is resolved by the rate-of-living theory. Large animals live longer than small animals because they produce free radicals at a slower rate. Cold-blooded animals live longer than warm-blooded animals for the same reason, and you can slow down your own aging by decreasing your rate of energy use. Couch potatoes of the world unite. Keep an eye on your diet and watch those exercising maniacs drop like flies. For once, biological complexity has collapsed into a satisfying simplicity.

You Knew It Was Too Good to Be True

Of course, life is never so simple. If it were, biology would be as simple as physics rather than the most complex of the sciences. One complication of the preceding scenario is that there were obvious exceptions to the metabolic rate–longevity trend. One of the most glaring is bats.

Bats, for all the unwarranted visceral disgust and bad press they generate, are arguably the most interesting of all mammals. The only mammals to have truly taken to the air, bats have been a raging success worldwide. In fact, more than one-fifth of all mammal species are bats. They are particularly numerous in the tropics, where they roost in caves, trees, tombs, and many other kinds of buildings. In some areas, so many large bats fill the sky at dusk that they are a danger to incoming aircraft and airline schedules have to be modified accordingly. Bats also seem to be especially fond of tropical movie houses, where you get double entertainment value if you enjoy, as I do, watching them course after moths in the projector's beam.

In the temperate northern latitudes of Europe, Asia, and North America, we have a pretty parochial vision of bats, because the only ones we commonly see are those mouse-sized insect eaters with grotesque faces specialized for receiving the sonar signals they broadcast to locate prey. These bats typically hibernate in caves, trees, barns, and attics during the winter when insects are not available.

But elsewhere, bats are much more diverse. In addition to insects, some bats eat fruit, others consume nectar and pollen, still others live on mice, fish, frogs, or in the case of vampires, pure blood. Many bats do not have sonar to help them locate prey; they find their food utilizing sight and smell and have faces more like foxes than Gothic gargoyles. Finally, and most important for understanding how bat longevity is connected to the rate-of-living theory, not all bats hibernate. Some, like birds, fly south in the winter, and others spend their entire lives in the tropics and neither hibernate nor migrate.

Northern parochialism misled gerontologists about bats for many years. Bats have been known for a long time to be exceptionally long-lived. For instance, the little brown bat of Canada and the United States is less than one-half the size of a mouse, yet instead of living a few months in nature, or a couple of years under the best captive conditions, it can live more than 30 years *in the wild*. Donald Griffin, the biologist who demonstrated in the 1930s, while still a student at Harvard, that some bats do in fact use sonar for navigation, once told me of a time when he was rechecking one of his caves many years after first putting leg bands on a large number of its little brown bats. His student assistant who was logging data in the field book suddenly looked up in amazement at the tiny animal he was holding and exclaimed, "This one is older than I am."

Though bats of extreme age have been identified, we still know little about the limits of the animals' longevity. Zoos don't frequently keep them; when they do, they keep large colonies and don't keep particularly good records of how long individual bats live. Virtually everything we know of bat longevity was discovered by accident. Researchers will band a cave of bats for some reason, and many years later someone (usually someone else) will happen to recheck the same colony and find some living animals that have leg bands. The interval between banding and rechecking defines what we know about bat longevity. Therefore, bats could be even more grossly long-lived than we now think.

But gerontologists, being mainly familiar with temperate-zone bats, and being somewhat blinded by an unyielding belief in the rate-of-living theory, refused until recently to acknowledge the truly exceptional nature of this life span. The assumption was that because bats hibernated in the winter, banking their metabolic fires in the process, and because they often became torpid, dropping their body temperature and metabolism while resting, even during times of the year when they were not hibernating, they were not, metabolically speaking, any longer-lived than other mammals. They just spent a considerable portion of their lives in a nonaging state of suspended animation, like insects embedded in ice. George Sacher even calculated, given some plausible assumptions about hibernation times, that bats were just mice with a bad case of narcolepsy.[49] In other words, if you considered the amount of their lives that bats spent with the low metabolism of torpor and hibernation, they expended about the same amount of energy per cell in a lifetime as did mice.

As it turns out, this isn't true. Or rather, it *is* true that many bats hibernate, and that during hibernation metabolism is greatly reduced. And bats do go into torpor when inactive, greatly slowing their metabolism. The key fact is that this reduced metabolic rate isn't responsible for their exceptional longevity. We know this because information is now coming in on tropical bats that do not hibernate and that become torpid less frequently. They are just as long-lived as their temperate-zone relatives. There has to be another explanation for bat longevity. They must have mechanisms for escaping, or dramatically slowing, free radical damage, if in fact free radical damage is a major part of aging.

There are other apparent exceptions to the rate-of-living idea, too. Sacher had information on the longevity of a few dozen mammals living in zoos. Today we have information on hundreds of species, and advances in zoo husbandry have made this information more reliable. The same general trend that Sacher identified still holds. Larger animals are longer-lived than small ones, as a rule. But there is no simple relationship between longevity and metabolic rate. The marsupials are one good example of this.

Marsupials consist of the 250 or so mammal species that typically bear their young in a pouch. Kangaroos, koalas, and opossums are some of the best-known marsupials. We usually associate marsupials only with Australia, but fully one-third of the species (opossums and their relatives) occur in South and Central America, though only one species, the Virginia opossum, occurs north of Mexico. Marsupials as a group separated from the rest of the mammals (called eutherians) more than 100 million years ago, and besides having pouches they are distinguished from the rest of mammals by having lower metabolic rates. Marsupials have metabolic rates only 70 to 80 percent as fast as the rates of eutherians of the same size. Therefore, according to the rate-of-living theory, marsupials should be longer-lived for their size. Opossums should live longer than cats. Kangaroos should live longer than chimpanzees or humans.

But they don't. Marsupials are shorter-lived. An opossum is old at two years, dead by three. Kangaroos rarely live into their twenties. In fact, no marsupial has ever been reported to live as long as a cat or a little brown bat, much less as long as a monkey, whale, or human. The longevity champions among marsupials appear to be the wombats, pig-like burrowing animals, which live into their mid-twenties. The largest kangaroos live almost this long, too.

And what about birds? Birds have *higher* metabolic rates than mammals, more than twice as high in some groups of birds. Therefore, according to the rate-of-living theory, they should be shorter-lived than mammals. Yet, metabolically speaking, birds might be the longevity champions of the animal kingdom. They live at least three times as long as same-sized mammals, if we ignore the bats. For instance, mouse-sized birds can easily live six or eight years in the wild, and 15 or 20 years in captivity. Reports of pet parrots, crows, ravens, and hawks older than 50 years are common. There have even been isolated, relatively well-authenticated reports of parrots living past the century mark.

And what about the dieting rats and mice? Are they really only stretching out their lifetime food allotment over a longer period by reducing their rate of metabolic expenditure per cell? When Sacher performed his calculations demonstrating that dieting rodents had reduced metabolism per cell, he needed to make a lot of assumptions about their food, its digestibility, and the manner in which rodents grew. The original paper he consulted to make these calculations never actually measured metabolism. Once the metabolism of food-restricted rats was actually measured in the mid-1980s, it turned out that although there was a brief dip in metabolic rate per cell at the beginning of their restriction, their metabolism soon rebounded and burned at least as fiercely per cell as it did in fully fed animals, maybe slightly more so.[50] The 20 to 40 percent life extension that rodent dieting produced was not, after all, purchased with reduced metabolism. Somehow these animals were capable of using their energy in a less damaging fashion.

What about Humans and Their Big Brains?

Marsupials, bats, birds, and food-restricted rodents were not the only obvious exceptions to the theory that there is a simple relationship between metabolism and longevity. There was also the awkward case of humans. Humans live more than four times as long as would be predicted from their body size. Not bat- or birdlike longevity, but nothing to sneeze at, either. Does this mean that we have a metabolic rate that is one-fourth the mammalian average? In a word, no. We have a somewhat higher-than-average metabolic rate for our body size, as do most of our primate relatives. Whereas Rubner's and Sacher's mammals lived long enough to expend only 200 to 300 kilocalories per gram of tissue per lifetime, humans expend about 800 kilocalories per lifetime.

One might have thought that this awkward fact, by itself, would have necessitated rethinking the rate-of-living notion. But it's impossible to overestimate our willingness to believe in our own uniqueness. If humans didn't fit the general pattern, that was to be expected. Instead of rethinking the entire idea, it was easier to try to account for what made humans so special.

Finding biologically unique features in humans is not so easy, though. For mammals, we are relatively hairless, but naked mole rats (perhaps the ugliest creatures ever created) make us look wooly by comparison, and no one has ever tried to relate longevity to hairlessness. Then there is the opposable thumb of which we are so proud. But contrary to popular belief, other mammals, such as opossums and tarsiers, also have opposable thumbs. Koalas even have six opposable thumbs—two on each forefoot and one on each hind foot. Finally, there is our large brain. Indeed, we do have an exceptionally, if not uniquely, large brain compared with the size of our body, about six times as large as that of an average mammal. Could it be that brain size, in addition to, or in concert with, energy use was a key to longevity?

The brain-size theory of aging arose in exactly this manner—out of anthropocentrism (human-centered thinking) by rate of living. And it wasn't an unreasonable theory. After all, most bodily functions are regulated to some extent by the brain. The larger the brain, one could imagine, the potentially more precise would be the control over these functions. Also, the human brain uses more energy than any other organ, so the brain-size and rate-of-living theories were in a certain sense complementary. If life is inherently destructive, but precise physiological regulation can somehow mitigate life's destructive forces, then brain size and metabolic rate could conceivably interact to produce the observable aging rate in any species.

Sacher developed this combination brain size–metabolic rate idea, which continues to have adherents today, in the late 1950s and 1960s using an elaborate analysis of mammalian trends in longevity, metabolic rate, and brain size. His analysis was purely correlational; that is, he determined that body weight (a reasonable index of metabolic rate) by itself accounted for about 60 percent of the variation among mammals in longevity. Brain size by itself accounted for considerably more—nearly 80 percent—of the variation, and body size plus brain size accounted for slightly more, about 85 percent, of the variation. Therefore, the ratio of brain size to body size must be a key to aging, right?

From this rather humble beginning, brain-size theory arose and developed. Pretty soon, some particularly enthusiastic scientists were claiming to be able to determine the life span of our extinct human ancestors by measuring the size of their brain cases. The somewhat awkward fact that Neanderthals had larger brains than modern humans, yet seldom if ever lived beyond 45 years old,[51] did not seem to discourage these researchers. By the mid-1970s, someone had even suggested that brain cells as they are growing and dividing might produce a vital substance that is necessary for bodily maintenance and repair. Once brain cells no longer divide (by about birth in mammals), the substance is no longer produced, but because animals with larger brains will have produced more of it, they will live longer.

But thinking back to the original source of the brain-size notion, it is important to note that it was based on nothing more than a correlation between brain size and longevity. And as every elementary statistics text warns, correlation does not necessarily mean causation. John Allen Paulos, author of the book *Innumeracy*, illustrates this point by noting a negative correlation between divorce and death rates in different regions of the United States. You could interpret this to mean that marriage is hazardous to your health. However, the real reason for the correlation is that both divorce and death rates correlate with something else—age. Older people are less likely to get divorced and more likely to die. The correlation is therefore inevitable, but divorce and death are not causally connected in any way.

What never occurred to Sacher, and what now appears to be true, was that brain size might simply be a more accurate or characteristic measure of body size than body weight, which is largely a function of diet. Sacher's original animal data came from the body weights of a mixture of lean, active animals measured in the wild and more obese animals measured in zoos. Some of his species, such as cattle, have been bred for centuries in the course of domestication for increased body fat. Brain size, which is not affected by degree of obesity, would therefore have been a better measure of typical body size. If body size and the aging rate are in fact correlated for whatever reason, is it any wonder that the correlation was stronger for brain weight than for body weight?

If there were something unique about the relationship between brain weight and longevity, then other organ weights might be expected to show a weaker correlation with longevity. In fact, heart, kidney, liver, and spleen weights all show at least as strong a relationship to longevity, as does brain weight.[52] Furthermore, although primates generally, and humans in particular, have exceptionally large brains and exceptionally great longevity, this particular combination is not especially general. Birds generally have brains that are equal to or smaller than they should be any particular body size than do mammals, yet they are about three times as long-lived. Amphibians, fish, and reptiles have only about one-thirtieth the size brain of mammals of similar size, but they are also longer lived. Even the longest-lived mammals for their body size, bats (LQs higher than 5), have brains that are no larger than average for mammals.

So despite the general trend for small mammals to be shorter-lived than large mammals, the fact that cold-blooded animals are often longer-lived than warm-blooded animals, and the fact that putting laboratory rodents on a Spartan diet prolongs life, a more detailed look at metabolism and life span doesn't provide much support for the rate-of-living theory.

Currently, the state of knowledge about metabolism and longevity is somewhat paradoxical. The rate-of-living theory is as dead as the proverbial doornail, but we still need to explain these puzzling general trends of aging and longevity in animals. On the other hand, free radicals are still very much in the game. We now know in more precise detail than ever how they are produced during normal metabolism and about the manifold damage they do to cells. So if the metabolic rate and the production of free radicals are not the keys to understanding aging, what is? To answer that question, we need to return to thinking about biological evolution and how it might have produced aging in the first place. If we can understand *why* aging occurs, the answer may provide more insight into *how* it occurs and what we might eventually do to mitigate it.

7

What Evolution Explains about Aging

*The Goate somewhat resembling the Sheepe in shape,
lives no longer, but is nimbler, and firmer flesh'd, and
should be therefore longer-liv'd, but lasciviousnesse
shortens his life. . . . The Cocke is leacherous, a
couragious fighter, and short-lived. . . . The Vulter
lives an hundred yeeres, Crowes also, and all ravenous
Birds feeding on flesh, are long-liv'd.*

FRANCIS BACON (1638)

Woody Guthrie, America's most famous folksinger, died of
Huntington's disease in 1967 at the age of 55. For at least 15
years prior to his death, he had shown signs of neurological deterioration—dizziness, jerky movements and facial tics, slurred speech, for-
getfulness, and periodic bursts of violent irrational anger. He was hos-
pitalized during his last 13 years, after he became too incapacited
to take care of himself or stay out of jail.

Huntington's disease is as cruel a fate as there is on this earth. Although ultimately fatal, the disease does not kill quickly or mercifully. Sufferers progressively lose more control over body and mind during the decade or two before they finally die. What's more, because the disease usually appears in early middle age and is inherited, on average, by half of the children of its victims, those destined to die of it have generally grown up with grim fatalism in a home devastated by the increasing emotional imbalance of a parent.

This was certainly the case with Guthrie, who inherited the disease from his mother, Nora. She began behaving erratically in her early thirties, and because of her unpredictability, Woody often stayed away from home even as a small boy. She eventually lurched so far out of emotional control that she chased her eight-year-old son George around the house with a knife because he refused to take an icy bath she had drawn for him. Mercifully, for the rest of the family at least, she was permanently confined to an asylum at age 39, just after she set her husband on fire with a kerosene lamp.

Huntington's disease, for all its horror, provides a crucial clue to solving the paradox of aging. That is, it points to how aging must inevitably evolve despite the enormous potential for self-repair inherent in animals. This was a point first noted by the geneticist J. B. S. Haldane.

It isn't particularly surprising that Haldane had this insight. He was arguably the most brilliant biologist of the century and inarguably the most original. A polymath par excellence, Haldane graduated from Oxford with first-class honors in both mathematics and classics before going on to a career in biology. He could be impossibly difficult and infuriating, likely to detonate in rage for no apparent reason and unwilling to compromise on the most trivial point. He lacked physical fear, having secretly enjoyed himself during the trench warfare of World War I. He also had a preference for experimenting on himself rather than on animals—drinking various acids to measure their effect on his breathing rate, or trying to understand the effects of oxygen deprivation by locking himself in an airtight chamber and reducing the oxygen level until he passed out. Haldane revolutionized or helped revolutionize scientific issues in evolution-

The folksinger Woody Guthrie (seated left) at about age 14 with his mother, Nora; father, Charley; and younger brother George. Nora had already been suffering from Huntington's disease for a number of years when this photograph was taken, and she probably has her arms locked behind her back to prevent their uncontrollable twitching. She transmitted the Huntington's disease gene to Woody, but not to George. (Courtesy of the Woody Guthrie Archives)

ary theory, genetics, biochemistry, and physiology. Politically, he was a communist for most of his life, and he left England at the height of his eminence for an academic chair in India, where he studied Hinduism and took to wearing robes in the laboratory. Haldane, writing mostly for blue-collar readers of *The Daily Worker*, also became our most incisive and lucid popular writer about biology since Charles Darwin.

What Haldane pointed out in passing was that Huntington's disease is probably so common because it doesn't appear until midlife in most individuals. That doesn't sound particularly profound, but its implications are. In a nutshell, those implications explain the ubiquity and paradox of aging.

Is Huntington's disease really common? In fact, it strikes about one person in 15,000 among those of European descent, which doesn't seem especially common to a nongeneticist. But Haldane realized that simple genes that cause fatal diseases should as a rule be rapidly eliminated by natural selection, because people having such diseases will generally leave few or no children.

Haldane's logic is captured in an old joke about how bearing children is hereditary. If your parents didn't have children, runs the joke, the chances are that you won't either. Although a bit ridiculous, this joke expresses the essence of natural selection. Any genetic trait that prevents reproduction, or reduces the number of offspring its bearers produce, should ultimately become vanishingly rare in a population, cropping up only due to recurrent mutations.

However, mutations are rare and random events, rather like meteorites falling out of the sky. We know from decades of genetic research that even a very frequent mutation of this sort should occur at most about once per many hundred thousand individuals. More commonly, these mutations should occur in only one in a few million people. One gene that we can compare to Huntington's to illustrate this point is the gene that causes progeria. Both the Huntington's disease gene and the progeria gene are dominant, meaning that individuals with only one copy of it inevitably develop the disease. Yet progeria, remember, occurs only once per 8 million births, which is

500 times less frequent than Huntington's disease. So by this comparison Huntington's disease *is* considerably more common than a geneticist might expect.

What Haldane realized was that if the effect of a defective gene was manifested only in midlife, as is true of the gene causing Huntington's disease, natural selection's power to remove it from a population would be diminished, because it would have a relatively small impact on lifetime reproduction. In fact, if the effect of the gene occurred very late—say, after reproduction had ended—then natural selection could not affect the fate of the gene at all, because it wouldn't influence the number of children its bearers had at all. From the standpoint of natural selection, the gene's effect would be neutral—that is, neither advantageous nor disadvantageous.

We can see how this works by looking at the Guthrie family. Nora Guthrie had five children before becoming incapacitated, and her son Woody had at least six (Woody was prone to extramarital flings, so we have to be cautious about stating his exact number of children). Guthrie genes, including the Huntington's disease gene, were therefore passed on at a substantial rate.

Although in retrospect Haldane's comment seems almost trivially obvious, it was in fact the highest sort of insight—the sort that, once you hear it, makes you wonder not whether it is true but why *you* didn't think of it first. Someone who may not have thought of it first, but realized its far-reaching implications for the evolution of aging, was the immunologist Peter Medawar, whose example of nonaging test tubes breaking in a busy laboratory was used earlier to clarify the difference between aging and longevity.

If a large part of genius is understanding that which others have merely observed, then Medawar was undoubtedly a genius. In his mind, Haldane's comment eventually expanded into an elaborate and precise evolutionary theory of aging, a theory that may in time dwarf the scientific contribution for which Medawar won his Nobel Prize— why the body rejects foreign tissues such as skin grafts and transplanted organs. His theory neatly resolves the paradox of aging and explains all the patterns in the natural history of aging that we have seen to date. It does so by specifying how natural selection's power to design a finely tuned and efficient body, by favoring or disfavoring helpful or harmful genes, respectively, gradually fades away as we get older.

Medawar was a slender six feet, four inches—a long stick of a man. His brilliance and energy were evident early in life, and they never flagged. For many years, he kept an active research laboratory running at the same time that he served as administrative head of England's largest medical establishment, the National Institute for Medical Research. During that period, he gave scores of lectures and wrote scientific papers and popular books. He was a professor by age 32, and a Nobel Prize-winner by 45. Even after a severe stroke at age 54 left him half blind, without use of his left arm, and able to walk only with the aid of a splint and cane, Medawar managed to keep active scientifically and wrote seven more books before his death in 1987.

To understand Medawar's idea, let's return to his analogy of the test tubes. Remember, we were imagining a medical laboratory full of clumsy, test-tube-using, scientists, who fail to replace any tubes they break. We also assumed that these test tubes don't age; that is, their accumulated nicks and scrapes don't make them more likely to break over time. Ultimately, of course, despite the fact that they don't age, all the test tubes will be broken regardless of the number we start with, and research in the lab will grind to a halt. If in one particular lab the chances of a test tube's being broken is 50 percent per year, then after one year only half of the original tubes will be left, of course. After five years, if tubes keep breaking at the same rate, only about 3 percent of them will remain, and after 10 years, there will be only about one tube in 1,000 left.

Now imagine that at the age of one year and once a year thereafter, those test tubes that aren't yet broken produce babies (tubelets? tubelings?). Notice that even without aging, most baby test tubes will have one-year-old parents, because they are the most common "adults" in the population. On the other hand, few tubelets will have five-year-olds as parents, and 10-year-old parents will be virtually unheard of.

We can now envision how Haldane's comment about Huntington's disease led Medawar to the idea that natural selection's power to influence the fate of genes gradually wanes as the age at which those genes have their effects increases. Imagine that half of the test tubes bear a suicide gene that causes them to break the moment they reach 10 years of age. Because there is practically no likelihood of a test tube reaching this age anyway, the tubes bearing this gene would produce on average about as many tubelets as would those without it. Thus, the gene would not be disfavored by natural selection. On the other hand, suppose half the tubes carry a fertility gene that doubles the number of tubelets born to tubes aged 10 years and older. Similarly, this gene would convey practically no evolutionary advantage, because there will be few, if any, tubes alive to benefit reproductively by it.

If the same suicide or fertility genes had their effect at the age of five years, then they would be slightly disadvantageous or advantageous, respectively, because they would have a 3 percent chance of affecting the number of tubelings produced in a life. If these genes had their effects at age two, then the fertility gene would be very advantageous, and the suicide gene very disadvantageous. That is, tubes bearing the fertility gene would leave many more tubelings than those lacking it, and soon, via the process of natural selection, the fertility gene would predominate in the population. Similarly, the suicide gene would be very disadvantageous from an evolutionary standpoint, because its bearers would leave relatively few tubelings, and natural selection would soon eliminate it.

Therefore, as a consequence of death due to accidents, the power of natural selection to influence the spread or elimination of genes becomes weaker and weaker as the age at which those genes have their effects increases. Note that the rate at which natural selection fades is related to the relative "hostility" of the environment. This fact will be important shortly. If the laboratory in which these hypothetical test tubes lived were occupied by fanatically careful scientists, so that only 5 percent of tubes were broken per year instead of 50 percent, then the fate of a gene that had its effect at 10 years of age *would* be strongly affected by natural selection, because about 60 percent of tubes would still be "alive" and reproducing at that age.

What does this scenario of waning natural selection on late-acting genes have to do with how aging evolves? The answer is that it leads directly to gradually increasing physical decay with age by two routes. Most obviously, any gene that affects the body *only late in life* is virtually immune to the forces of natural selection, so that even if the gene caused something catastrophically harmful to occur, natural selection would be unable to root it out. It would become more common or less common due to chance alone. Because the vast majority of mutations are harmful—that is, they are as unlikely to make you healthier as a random change in your car engine's timing is likely to make it run better—mutations causing late-acting harmful effects have accumulated in the human genome over hundreds and thousands of generations the way flotsam and jetsam accumulate on a beach with time. And because mutations occur in all parts of the genome, this "mutation accumulation" should lead to harmful effects not in just one or two organs but throughout the body. So it's not just your lungs or liver or eyesight that goes, but a bit of everything.

One example of exactly the sort of gene Medawar imagined may be the previously mentioned ApoE gene. Besides influencing cholesterol levels and the development of atherosclerosis late in life, it apparently also plays a role in the development of Alzheimer's disease.[53] ApoE is a protein secreted mainly in the liver, and is involved in removing cholesterol from the blood. There are three forms—ε2, ε3, and ε4—of the gene that produces ApoE; each of these forms produces a slightly different version of the protein. Each of these versions is also associated with differing blood cholesterol levels. Typically, the ε4 form of the gene is the most harmful, being associated with the highest cholesterol levels. It is particularly common in Finland, the European country with the highest cholesterol levels and heart-disease frequency, and it is particularly uncommon in China and Japan, countries with much less heart disease than the United States or Europe. Alzheimer's disease is also more likely to strike those people with one or two copies of ε4 and strike them at an earlier age than the rest of the population. All in all, ε4 is just bad news.

By contrast, the ε2 gene form seems relatively beneficial. It is associated with lower cholesterol levels. Not surprisingly, it tends to be more common where heart disease is less common, and vice versa. There is also gradually accumulating evidence that ε2 may protect against Alzheimer's disease. People with one copy of ε2 are less likely to get Alzheimer's disease than the population at large.

A recent study of French centenarians found that these forms of the gene changed with age, just as you might expect. That is, those people who had made it to 100 years of age had about half the proportion of ε4 and twice the proportion of ε2 as the French population as a whole. Thus, all of the evidence suggests that ε2 helps lead to longer life and ε4 to shorter life, with ε3 falling somewhere in between.

Given this information, a naive evolutionist would assume that the most common form of the gene would be ε2, because natural selection would have favored its apparent health advantages. By the same token, they would assume that ε4 would be the least common because of its health disadvantages, again with ε3 somewhere in the middle. But the naive evolutionist would be dead wrong. In fact, although populations vary somewhat, in all populations studied so far ε3 is easily the most common form, making up 60 to 80 percent of all ApoE genes. The unhealthy ε4 gene is next most common (7 to 24 percent), and the healthy ε2 gene is least common of all (5 to 11 percent).

How can this pattern be explained? At this point, we are not entirely sure, but as Medawar would no doubt have pointed out, the various ApoE forms seem to affect only illnesses of late life, which means that the frequency of the various forms of the gene may not be due to the effects of natural selection at all. It may be due mainly to chance. The ancestral gene is ε4 (we know this because it is the only form found in monkeys and apes), but at some time since we diverged from the rest of the primates, mutations have occurred, resulting in the ε2 and ε3 forms. It is also clear that the known health benefits associated with these newer forms of the gene affect diseases of later life. Therefore, the form of the ApoE gene that one carries will have no reproductive impact, so far as we know. Consequently, these gene forms are immune to being favored or disfavored by natural selection. They should drift erratically in frequency due to the chance occurrence of their being passed on. What is puzzling is why ε3 should be so consistently common in all cultures and countries. There are obviously aspects of the ApoE gene that are not yet entirely understood.

How common are late-acting genes like this? The short answer is that no one knows. No one even has an educated guess. But as we map and explore the human genome over the next few years, we should begin to learn which particular genes share this feature of affecting health only at late ages.

The second evolutionary route to aging has to do with the fact that there are some genes that have multiple effects. Many, maybe most, genes do. If a gene has a beneficial effect early in life, when natural selection is powerful, but a detrimental effect later on when it is weak, natural selection will actually favor that gene even if the late-life effect is catastrophic. For example, a new genetic mutation might benefit early-life reproduction by, say, hastening sexual maturity or attractiveness to the opposite sex, but decrease the chances for later life survival by causing heart disease or cancer (as we shall see later, these effects are not purely hypothetical).

This is all very fine in the abstract, but specifically what sorts of genes of this type might really exist? George Williams, the evolutionary biologist who fleshed out this idea, imagined for illustrative purposes a gene that hastened the hardening of bone early in life, thus adding early strength to the skeleton but resulting in calcification within artery walls later in life, producing arteriosclerosis.

One way to think somewhat metaphorically about how these genes work is by imagining that they mediate trade-offs in physiological resources. That is, at any given time we have only a finite quantity of resources to devote to our essential bodily tasks of survival and reproduction. If we shift more resources into one task—say, reproduction—fewer resources will be available for other functions, such as preventing disease or damage from free radicals. So let's imagine a situation in which in order to prepare our bodies for reproduction, and even during the reproductive process itself, we have to suppress temporarily our immune system or the enzymes that repair damage to our DNA. The effects of this decreased immunity or genetic repair may not show up until years later. But if even a small bit of genetic damage went unrepaired at that time, changing a normal cell to a precancerous cell, for instance, and if the immune system did not detect and destroy that cell, then we might later suffer serious consequences.

There is ample evidence that there is, in fact, a longevity cost to reproduction. For instance, testosterone, the male reproductive hormone, suppresses the immune system in exactly this fashion. It also accelerates deterioration in arterial walls, and is involved in the development of prostate cancer, to name a few of its many charms. Consequently, if you castrate a male guinea pig, he will fight off infections much better,[54] and if you castrate an adolescent dog, he will not get prostate cancer. Testosterone, in other words, may reduce life expectancy by means other than simple teen-age dementia.

We know this is true for *Antechinus*, the marsupial mouse mentioned earlier in which all males die from an abrupt immune-system collapse after their frenetic two-week-long breeding period. Castrate these males (or simply isolate them from females) and their immune systems are fine. They live blithely on for several more years, just as a normal house mouse would.

Believe it or not, this may also be true for humans. We know this because of some disturbing, but revealing, data excavated by the physician James Hamilton in the 1960s. He turned up longevity records for inmates incarcerated in an unnamed institution for the mentally retarded in Kansas. A number of years ago, it was common practice to castrate mentally retarded people living in institutions, especially those who posed behavioral problems. About 300 of the men whose records Hamilton found had been castrated. These were compared with more than 700 men living in the same institution who had not. What was most surprising was not that the castrated men lived longer, but how large the difference was. The castrated men, on average, lived almost 14 years longer than did the uncastrated inmates.[55] Moreover, the life-extending effects of castration were greater the younger the inmates were when the surgery was performed. Although castration may be one remedy for some of the causes of aging, it probably will not catch on as a popular remedy.

Women's reproductive hormones have some detrimental long-term effects, too. High doses of estrogen, as occur during pregnancy, suppress the immune system just as testosterone does (although lower estrogen doses can stimulate immunity). Also, some relatively common women's health problems can be traced directly to estrogen's effects.

Breast cancer is the most common cancer in female dogs, occurring at two to three times the rate in humans.[56] However, if a female dog is spayed before puberty, depriving her of her main source of estrogen, she is less than 1 percent as likely to develop breast cancer as is an intact female. If she is spayed after her second reproductive cycle, her risk climbs to 26 percent that of unaltered females. Although breast cancer is less common in cats than dogs, the same pattern exists, with spayed females having about one-seventh the risk of ultimately contracting the disease. In humans, of course, breast cancer is very common, now striking about one out of 10 women in the United States and Europe. As in cats and dogs, in humans there is little question that breast cancer is due primarily to one's cumulative lifetime exposure to reproductive hormones, a topic discussed in detail later on.

These reproductive "costs" are not news to everyone. Veterinarians have known for years that their "fixed" animals are the longest-lived, whether they are males or females. The pathologist Roderick Bronson—the only person I know who, with his outspoken opinions and deadpan humor, can keep an audience awake while viewing carousel after carousel of tissue slides—examined the autopsy records of hundreds of dogs and cats from a large Boston veterinary hospital and found that male dogs and cats that had been castrated lived more than three years longer than did their intact counterparts. Spayed females of both species lived four to five years longer.[57]

Some sort of vague relationship between aging and sex or reproduction has been speculated about for hundreds of years. Ancient folk wisdom said that aging men could be rejuvenated by lying between two (female) virgins. In the late nineteenth century, the world-renowned physician Charles-Édouard Brown-Séquard acquired the belief that injections of extracts from the testicles of various animals would relieve many of the physical ailments of aging. To test his beliefs, Brown-Séquard followed a fine, but alas vanishing, medical tradition of experimenting on himself. At age 72, he began injecting himself with extracts from dog and guinea pig testicles. He observed that he suddenly felt stronger and more mentally and physically vigorous, and even his control over bladder and bowels had improved. The phenomenon known as the placebo effect was not understood in those days, so there soon followed an embarrassing period of medicine in which extracts from testicles and other organs were used widely to treat a variety of diseases, from mental senility to tuberculosis. Such fashionable quackery reappears with depressing regularity as treatments for aging, as we shall see later. Indeed, it continues today, thanks largely to scientific illiteracy combined with wishful thinking.

The effect of reproductive hormones on subsequent aging is only one of a host of such physiological trade-offs. It is beneficial in the short-term, for instance, to have certain white blood cells packed with damaging free radicals. These free radicals kill invading bacteria when the white cells burst. But in doing so, radicals are inevitably released into the bloodstream, where they can damage blood-vessel walls. A bit later, I will examine some of the most important of these trade-offs to demonstrate how they might allow us to circumvent some aging processes. For now, it is important to remember that the evolutionary theory of aging describes *genetic* processes. Presumably, the physiological effects mentioned so far are underlain by genetic factors, but it would be even stronger support for the theory to deterimine that *genes* altering reproduction also modulate aging and longevity.

There should be a mountain of information available on this issue. Humans have been selectively breeding—that is, genetically altering—various domestic animals such as chickens, pigs, and cattle for early and copious reproduction for hundreds of years. If there is a genetic trade-off between early reproduction and longevity, as evolutionary aging theory suggests, then comparing aging of these highly productive genetic breeds with that of their unselected ancestors should reveal it. Alas, this information doesn't exist. People in the business of animal breeding have no patience with animals that are beginning to slow down. As soon as reproduction begins to falter, animals are, shall we say, "discarded." So we have to look elsewhere for genetic evidence of trade-offs between early reproduction and rapid aging.

The most elegant experimental science addressing this question comes from work with *Drosophila*, the fruit flies that buzz around the bananas in your house in the summer, and also the old standby of laboratory geneticists. Possibly even less charismatic than rats and mice, *Drosophila* are wonderfully designed for studying genetic change. They are small (hundreds of thousands of them can be kept in a small laboratory), easily bred, have a new generation every two weeks, and live only a month or so.

The earliest work on the genetics of aging using *Drosophila* was done by a one-time student of Haldane's, John Maynard Smith. Like his mentor, Maynard Smith came from a background in the quantitative sciences. One of the most pleasant men you will meet, with straggly shoulder-length hair and Coke-bottle-thick spectacles, he has a cackling laugh and a propensity for making you feel as if you are a major contributor to the endless stream of new ideas that pour from him. Having originally trained as an engineer, Maynard Smith spent World War II designing fighter planes before developing an interest in evolutionary biology and going to work with Haldane. They were political soulmates as well. Maynard Smith was a Marxist until the invasion of Czechoslovakia in 1968, and because of his political history, he was not able to obtain a visa to visit the United States until the past few years. This study of aging, as it turned out, was a passing fancy, just as Marxism had been. Maynard Smith made his real mark in evolutionary theory, particularly in the application of the theory of games to biological problems.[58]

But in the 1950s, he was very much fascinated by the relationship between longevity and reproduction. He compared longevity changes using several methods of halting fruit-fly reproduction. First he simply kept females isolated from males. When he did so, they laid fewer eggs, naturally, and also lived longer. Then he tried sterilizing females by rearing them at temperatures too high for normal egg development. These females also lived longer. Finally, he wondered whether genetically reduced reproduction would also extend life, so he bred a group of fruit flies in which all females carried a genetic mutation that inhibited development of their ovaries. They lived the longest of all.

In perhaps the most dramatic genetic experiment supporting the evolutionary theory of aging, fruit flies were again used. In this case, the experiment was an evolutionary one, the goal of which was to create a race of long-lived flies. The experimenter's name was Michael Rose.

Rose is sharp featured, sharp tongued, and, like Miniver Cheevy, imperially slim. Whereas Maynard Smith puts forth his ideas with a touch of diffidence, Rose puts them forth with defiance. If he has any self-doubts, they are afraid to show themselves. As someone once said about Winston Churchill, "I wish I were as certain about anything as he is about everything."

Rose's experiment was simple and elegant. He would artificially slow down the rate at which natural selection waned with age. To do this, for generation after generation of fruit flies, he discarded all eggs produced before the fruit flies had reached three weeks of age and therefore only allowed eggs produced relatively late in life to form the next generation. Successful reproduction, therefore, required surviving and still reproducing well at an age at which a large fraction of normal flies had already died. Genes that had detrimental effects on survival at any time before three weeks of age, when reproduction began, would suddenly become highly disadvantageous (instead of mildly disadvantageous) and selection should eliminate them. On the other hand, genes that favored longer survival at the expense of, say, early reproduction would suddenly be much more advantageous. In this way, by making reproductive fitness at an older age very important, genetic changes over the generations ought to eventually produce longer-lived flies.

To clarify Rose's logic, think of the flies in terms of our previous nonaging test-tube population, in which reproduction began at one year of age, and 50 percent of the test tubes "died" each year. Remember that at this high death rate, only 3 percent of the test tubes were still "alive" at age five and fewer than one in 1,000 at age 10, so natural selection's power waned quickly because test tubes were unlikely to survive to advanced ages. What if we delayed the age of first reproduction from one year to five years? Now, even though only 3 percent of individual test tubes live that long, those survivors produce *all* of the next generation. As a consequence—unlike the earlier case in which genes affecting test-tube fitness at age five or later only minimally influenced the total offspring left—now the genes affecting survival at age five will be enormously influential in determining the extent of individuals' contribution to the next generation. That is, instead of being virtually neutral, genes with effects at age five suddenly become enormously important. If they are detrimental, they will be quickly eliminated from the population, and if they are beneficial, they will quickly spread. The net effect is that the aging rate should slow as more and more genes favorable at age five and beyond accumulate and more and more genes detrimental at age five and beyond disappear.

Rose expected to be able to create longer-lived flies using this method, and he did. After 12 generations, his late-reproducing flies lived 10 percent longer than normal flies. In later experiments, he progressively increased the age at which reproduction began until his flies started reproducing at an age when none of the original flies would have been alive. He now has flies that live about twice as long as his original ones. The evolutionary theory of aging allowed him to put the brakes on aging in fruit flies within a relatively short time.[59]

Methusaleh Opossums

It was at about this time—in the mid-1980s—that I became interested in the biology of aging. Maybe it was a life change. Earlier in my career, as a graduate student and postdoctoral researcher, I'd studied sex—sex in spiders, sex in birds, sex in mammals, sex in beetles. I prefer to think of my new direction as a shift of focus, not an actual change of interest. I also prefer to think of it as due not to my own encroaching decrepitude, but to scientific serendipity. I'd discovered quite by accident that opossums aged much faster than anyone suspected, and I became determined to understand why.

It was, in fact, serendipitous that I was even studying opossums. I had been living in the dust and mud of the Venezuelan savanna at a small biological field station in order to investigate obscure questions about the mating behavior (what else?) of a tropical wren. A friend of mine, Mel Sunquist of the Smithsonian Institution, lived at the same field station and was trying to study crab-eating foxes.

Sunquist is a rangy, grizzled veteran field biologist who has worked on most of the earth's continents. If it walks, crawls, runs, climbs, digs, or swims, he has probably slapped a radio collar on it at some time. He also takes a continuing glee in his work that I envy. When he releases a newly collared animal, he's likely to sing out, "Here comes number six and she's *on the air!*" as it ambles away into the bush.

I say he was *trying* to study foxes, because for once he was having little luck. His problem was that he couldn't trap many foxes. The reason was that his traps were filling up with opossums each night instead. Always a wiseacre, I piped up, "Forget foxes; check out those opossums. There must be *something* interesting about them." What that something was, of course, I hadn't the faintest idea.

Soon Sunquist and I had radio-collared several dozen opossums, and we were recapturing them once a month to study their reproductive behavior. We were astonished to discover almost immediately that an apparently vigorous 18-month-old female in seemingly fine health would several months later have deteriorated into a feeble crone. She might have developed cataracts, be tottering along with an arthritic gait, be suddenly losing fur, and be festering with parasites. We never found opossums older than about two years. They seemed to experience greatly accelerated aging. What was going on?

Having read Rose's paper on the creation of long-lived flies, I wanted very much to determine whether the same evolutionary theory applied to mammals. After all, mammal biologists are archchauvinists. They think mammals are uniquely complex and uniquely interesting. And because humans are mammals, you might expect that no one would take this evolutionary theory seriously with respect to its potential for understanding—and perhaps leading to medical treatments for—human aging until it had been shown to apply to another mammal, at least. Opossums, with their accelerated aging, seemed like an excellent subject for such a study.

But mammals make terrible subjects for following genetic changes. Even rapidly aging ones live too long and take too long to reach puberty so that you can breed them. Trying to repeat Rose's selection experiments, even with very short-lived, fast-breeding mammals such as mice, would take more than a decade and be enormously expensive. I decided instead to look for an island off the southeastern coast of the United States, where I suspected nature itself might have already done the experiment for me—that is, evolved a long-lived population of opossums.

My logic was this. Opossums and other small mammals generally die in nature from predators or cold temperatures. Cold temperatures would not usually be a problem for animals the size of an opossum (about the same size as a house cat) in the southeastern United States. Therefore, their most likely cause of death was predators. Because they are slow moving and not terribly well armed with claws, teeth, brains, or agility, opossums will be killed by nearly every type of predator—owls, coyotes, wolves, feral dogs, cougars, bobcats, and, perhaps their worst enemies, automobiles. Okay, cars have been opossum predators for less than a century, so in evolutionary terms they probably have not been important. Hawks are also not much of a problem because opossums are night creatures. Hawks hunt by day and sleep, very sensibly, by night.

My hypothesis for opossums' rapid aging was that, given all these predators, their environment was a very hazardous one. Remember that in an extremely hazardous environment, the high rate of accidental deaths will make natural selection fade rapidly with age. In these circumstances, an animal's time horizon is very short. If a predator is likely to kill you in the next few weeks or months, it makes little sense to waste resources on a long-lasting, effective immune system or array of free-radical defenses. It is better evolutionarily to reproduce copiously, and the sooner the better. This argument makes strong intuitive sense, as well. The most ardent antismoking advocate probably would not waste his time trying to convince soldiers in combat or death-row inmates that smoking is bad for their health.

Islands were of particular interest because they are renowned among ecologists for their lack of predators. As a consequence, island animals are typically characterized by a lack of fear of such things as dogs, cats, or humans. This is why mariners formerly stopped at islands during long ocean voyages. They could saunter among the birds or turtles or lizards or whatever and casually club them to death for fresh meat. Even today, on the Galápagos Islands of Ecuador, I've had Galápagos mockingbirds perch on my shoes to peck for seeds in the eyelets. I've also seen herds of Galápagos fur seals barely bother to open an eye as gaggles of tourists walked among them chatting loudly and snapping photographs. This lack of defensive fear is largely responsible for the large number of extinctions of island animals in the past few centuries.

The reason that predators are rare or nonexistent on islands is that they require a continuous supply of prey to exist, and most island populations of prey are too small and produce too few young to support a sustainable population of predators. By this logic, island environments should be less hazardous than mainlands, and, if the theory is correct, animals living on islands should evolve slower aging over an evolutionarily significant amount of time.

I began canvassing the southeastern United States—opossum heaven, as it were—for islands that had several specific characteristics. First, they had to have opossums. This would make an opossum study considerably easier. Second, the islands had to be distant enough from the nearest mainland that opossums were unlikely to swim there. If opossums could swim there, then mainland opossum genes and island genes would be constantly mixing, and no specific genetic adaptation to the island environment would occur. Finally, the island had to be old enough, meaning it had been separated from the mainland for long enough that evolutionary change could occur.

After much time searching, I found my island—Sapelo Island, Georgia. Sapelo is a low-lying barrier island, about five miles of salt marsh and open ocean distant from Meridian on the mainland coast. Shaped a bit like a dagger, it consists of more than 10,000 acres of "high ground," where ocean tides never reach, and about half again that amount of grassy salt marsh. It is occupied by very small deer, large rattlesnakes, and more ticks than I've seen anywhere. Geological information indicates that the island is about 4,000 years old and was occupied by Indians perhaps from the very first. Most of the island today is a state-owned wildlife refuge, but there is also a small village, Hog Hammock, and the University of Georgia Marine Institute. Hog Hammock is the home of about 70 Gullah-speaking descendants of the slaves of Thomas Spalding. Spalding and his heirs owned Sapelo during the nineteenth century. In this century, the island was owned by Howard Coffin, one of the founders of the Hudson Motor Company, and Richard J. Reynolds, an heir of the original tobacco magnate. Reynolds is ultimately responsible for the formation both of the Marine Institute (its laboratories are housed in what used to be his dairy barn) and the wildlife refuge, but today the entire island is state owned except for the 434 acres that make up Hog Hammock.[60]

So Sapelo seemed to be just what I sought. Because of its connection with the University of Georgia, there had been ecological surveys of the animals there. From these surveys I knew there were no pumas, foxes, bobcats, or other mammal opossum predators on the island. I also knew there *were* opossums. Furthermore, there was excellent reason to suspect that these opossums were indeed isolated from mainland contamination. At the end of the Civil War, opossums, always a popular food in the South, were hunted to extinction on all of the Georgia barrier islands except Sapelo. Since then, opossums haven't recolonized any islands not connected to the mainland by bridge. Therefore, it seemed improbable that opossums would swim across the five miles of ocean when they didn't much like to be in water in the first place. All my requirements were met. It remained only to determine whether Sapelo opossums indeed aged more slowly than their mainland relatives.

I soon set up shop in a small travel trailer in Magnolia Circle, a picturesque clearing amid a forest of massive oaks festooned with Spanish moss. The Marine Institute lent me a pickup truck to navigate the welter of sand and the gravel roads on the island, and I began placing radio collars on all the female opossums I could catch. I concentrated on females, because if they had young in their pouches, I could mark them too, so that if I later recaptured them I would know exactly how old they were.

Opossums, like other marsupials or pouched mammals, bear young that are very tiny, specifically about the size of a large ant, and only partially developed. At birth, opossums are blind, deaf, hairless, and do not even have fully formed hind limbs. They have to make a perilous half-inch journey between the birth opening and the pouch by grasping their mother's fur with their well-developed front feet and pulling themselves up into the pouch. Many don't make it. So although 20 or 30 young may be born, there are only 13 nipples in the pouch, and typically only five to 10 pups are weaned. Developmental biologists have studied at length the age at which eyes open, teeth erupt, or fur or genitals appear in opossum pouch young. Therefore, I had a good set of markers with which to determine the age of pups in a mother's pouch whenever I caught a female with a litter.

Immediately I could tell that these animals were unusual. The first one I caught stumbled across the road in the middle of the day, and I leapt out of the truck and nabbed it by hand. I'd never seen an opossum walking around at midday before. I also noted that unlike animals on the mainland that slept only in underground burrows, Sapelo opossums were most often found lying on top of the ground, perhaps snoozing under a palmetto leaf. They had indeed lost some of the adaptations mainland animals are equipped with in order to avoid predators.

By the end of my first year of work, I knew that another expectation of evolutionary aging theory had been met. These opossums, in addition to losing some fear of predators, had also reduced their reproductive rate relative to their mainland relatives. Mainland animals generally have litters of six to nine pups, whereas Sapelo animals typically had four to six pups at a time.

As the longevity results began to come in, I became more and more excited. In two years, it grew obvious that Sapelo opossums were longer-lived. They averaged about 25 percent greater longevity than that of opossums on the mainland, and maximum longevity was some 50 percent greater. Even more convincing was the fact that the rate at which the chance of dying increased with age (aging in its purest form) was only about half as fast on the island as on the mainland.

What's more, the reproductive systems of Sapelo opossums aged more slowly. Mainland opossums rarely live into a second reproductive season. The few that do manage to make it to a second season are more frequently sterile, more frequently abort partially developed litters, and have more slowly developing pouch young than in their first season. None of this was true on the island, where more than half of the animals lived into a second breeding season, and they were just as fecund and fertile as in their first season.

I also discovered that Sapelo opossums' tendons aged more slowly. Looking into tendon aging wasn't just a lark, something to occupy the time when I wasn't picking ticks. What happens to tendons reveals a great deal about what is happening throughout the body.

Tendons are composed almost entirely of collagen, the most common protein in our bodies. Collagen provides our structure and support. It is the major component of teeth, bones, cartilage, blood vessels, tendons, ligaments, and skin.

Collagen, unlike most molecules of our bodies, has a long life. It may seem suprising that *all* our molecules aren't as long-lived as we are, but they aren't because of molecular turnover. The essence of molecular turnover is captured by the patter of an ax-juggling street performer I once heard. It went, "You may not believe it, but these very same axes once belonged to Genghis Khan." (Juggle, juggle) "Of course, we had to put on new handles." (Juggle, juggle) "Then we had to change the heads. . . ."

Most of the molecules of which we are composed turn over; old parts are continually being replaced by new, identical parts, just like Genghis Khan's axes. That is, they are broken down and reformed from different atoms—usually new atoms that we have acquired from our food. So even though we develop no new nerve cells in our brains after birth, for instance, brain-cell components such as proteins and fats *are* constantly being broken down and replaced. This doesn't mean that brain cells don't age—they do, but not because their component molecules are old and decrepit. In collagen and a few other tissues, molecular turnover is practically nil. Once mature collagen forms, it is chemically inert. Its molecules and atoms turn over ever so slowly, so the molecule itself ages.

Normal young collagen is strong, flexible, and relatively inelastic—something like an automobile tow strap. As it ages, it remains just as strong, or becomes stronger even, but it gradually loses its flexibility. And because joints are strapped together with collagenous ligaments and muscles connected to joints by tendons, you can see why 80-year-olds do not make good gymnasts.

The reason that collagen fibers become less flexible with time is that chemical crosslinks gradually appear and accumulate within the collagen fibers. Think of three steel chains twisted around one another to form one superstrong chain. This is a fair representation of a collagen molecule. Now imagine that we begin welding struts between adjacent links in each chain and also between the individual chains themselves. It is easy to see that as more and more struts are added, the chain's flexibility will gradually be reduced. This is essentially what happens to aging collagen. We will look at why this occurs a bit later. For now, you can imagine that if this process is occurring in critical places such as artery walls, it isn't helpful to cardiovascular function, and if it is occurring in joints it probably isn't helpful for agility. In fact, tendon fibers from long-lived animals age more slowly than do those from short-lived animals, and laboratory experiments in which aging was retarded also found that tendon aging was slowed. Tendon aging seems to be a general marker for how rapidly animals age.

Simple laboratory methods exist by which to measure tendon aging, so each time I recaptured my opossums, I would surgically remove a few tendon fibers from their tails. This was easily done in a couple of minutes. After suturing shut the inch-long incision, I would release the opossum, and a few months later when I caught the animals again I would repeat the procedure. In this way, I could track changes in the tendons of individual animals.

There was no reason to think that this procedure damaged them in any meaningful way. I removed only a small percentage of the fibers in each tendon bundle, and the incisions healed so cleanly that I often couldn't identify where the previous one had been.

Months later, back in my laboratory at Harvard, I was gratified to discover that indeed there was clearly slower aging in the tendons of Sapelo animals than in the mainland animals with which I was comparing them. Thus, by every measure I had examined—mortality rates, reproductive decline, and tendon aging—these island opossums were aging more slowly, just as the theory suggested they should.[61]

Now A Lot of Things Make Sense!

The logic of this evolutionary theory of aging seems sound, and abundant evidence from the laboratory and field supports it. But the power of a scientific theory is in the scope of information it explains. Its utility is in suggesting new and novel research directions. Before examining the theory's utility, it may be worthwhile to examine in detail its scope. In short, can it explain those general patterns of longevity mentioned earlier, and does it lead to the discovery of new patterns?

First, let's revisit semelparity, the habit of breeding in one suicidal burst, as do Pacific salmon, some octopuses, and the marsupial *Antechinus*. Semelparity has always been difficult to explain, except in erroneous group-selection terms. The rate-of-living theory, relying only on lifetime energy expenditure, has had no success whatsoever in explaining the phenomenon. Death among semelparous animals is associated with reproduction, not with some energy tank that has finally hit "empty."

Evolutionary aging theory explains semelparity quite simply as an adaptation to ecological circumstances in which attaining an appropriate reproductive situation is unusually difficult and not likely to be achieved more than once due to the environmental hazards involved. Salmon, of course, have been filmed many times as they persistently fling themselves against raging cataracts during their arduous, generally doomed attempts to return upstream to the river of their birth. This hazardous migration from birth stream to the ocean, and the later return to the birth stream, is achieved once by a tiny fraction of fish and is very unlikely to be successfully repeated. In such a situation, it becomes advantageous to invest every bit of available energy in reproduction once successful migration is achieved. Not surprisingly then, semelparous animals such as salmon and lampreys reabsorb much of their fat, muscle, and immune and digestive systems during reproduction, leaving no possibility of subsequent survival. Similarly, *Antechinus* is a tiny mammal living in a highly seasonal environment in which reproduction can occur only during a small window of time. Small mammals are subject to manifold hazards. Mice, as mentioned, live only three to four months on average in nature before something—an owl, a cat, a flooded burrow—gets them. If *Antechinus* manages to survive the 11.5 months until its first breeding season, it wouldn't be evolutionarily prudent to hold back reproductively in the hopes of surviving an additional year to yet another season.

A glaring difference between Pacific salmon and *Antechinus* is that in salmon both sexes die soon after spawning, whereas in *Antechinus* only the males die inevitably after reproduction. The evolutionary explanation is straightforward. In salmon, no parental care of eggs or small fry is required for the young to survive, so successful reproduction only necessitates survival until spawning. Survival much beyond that would be at the expense of additional eggs or sperm that might have been produced. This is also true of male *Antechinus*, which, like most male mammals, are little more than mobile sperm sources with no parental role to play at all. So it makes sense to give your all to the one and only mating season you are ever going to experience. Male *Antechinus* thus die of maleness and stress, probably in that order. During their two-week frenzy of mating, male reproductive hormones (androgens) increase more than eightfold, prostate size increases more than fiftyfold, and active stress hormone levels shoot through the roof (increase tenfold). Both androgens and stress hormones suppress the immune system, so the complete immune-system collapse by which they die is hardly surprising.

Females are in a very different situation. Their continued existence is crucial to the survival of their young. Specifically, female *Antechinus* feed and protect their young for about four months following birth. This is a considerable period of environmental uncertainty during which food may suddenly become scarce, unseasonable weather may erupt, and predators lurk around every bush. In such uncertain circumstances, a prudent parent needs to reserve some margin of energy for unexpected environmental calamities. Thus, females do not reproduce suicidally. They generally survive their first reproductive event, and very, very occasionally even survive to reproduce again the following year.

In addition to semelparity, another general pattern that any aging theory must explain is the relationship between body size and longevity. Can evolutionary theory do better than the rate-of-living idea did in explaining this relationship? It can. Evolutionary theory explains the relationship simply, by noting that large animals in general are less vulnerable to death from environmental accidents. Less vulnerability will lead to the evolution of slower aging.

In general, large animals should be vulnerable to fewer predators. Opossums may be preyed upon by owls, coyotes, wolves, feral dogs, cougars, and bobcats, but horses aren't. Furthermore, by virtue of their slower metabolic rate, large animals will be able to survive longer without food and water than small animals, and thus will be able to live through more environmental catastrophes. Finally, large animals can better resist climatic extremes for the same reason that a large block of ice takes longer to melt than an ice cube, and a large pot of water takes longer to boil than a small pot. The reason is what is called thermal inertia and has to do with the fact that large objects have relatively smaller surfaces compared with their volume than do small objects, and heat is transferred across surfaces. Chop your block of ice up into hundreds of cubes, spread them across the ground, and they will all melt as fast as your single cube. Combining less vulnerability to predators, food and water shortages, and climatic extremes makes a considerably less hazardous environment. Large animals should therefore age more slowly for the same reason my island opossums aged more slowly—a long history of exceptional safety.

The rate-of-living theory, in all fairness, is just as successful at explaining this very large (and very loose) trend. Where it fails is in explaining the smaller-scale patterns within the larger picture: why birds and bats age so slowly, for instance, and opossums so rapidly.

Some of these answers will be apparent by now. Birds and bats have fewer environmental hazards because of their gift of flight. They can fly away from ground-based potential predators, such as cats and coyotes. They can also quickly move large distances from an area of food shortage to one of food abundance. They don't have to worry about their burrows flooding. It has even been documented that flying birds and mammals (i.e., bats) have lower mortality rates in nature than their nonflying relatives. It makes perfect sense, then, according to evolutionary theory, that in the protected life of captivity flying animals should age more slowly.

At this point, it is important to remember that although the rate-of-living theory fails to explain many patterns of aging, it does lead to the important finding that metabolic processes, such as the production of free radicals, are damaging to animals. Is there a way that the evolutionary theory of aging can explain why, despite the damaging effects of metabolism, some animals such as birds and bats seem almost impervious to its damage?

As I said earlier, animals—in fact, all living organisms—can almost be defined as things that are capable of repairing damage to themselves. In fact, defense systems have evolved to fight free-radical damage as well as the other destructive processes of life. Some animals—those likely to live a long time—have evolved effective and elaborate defenses against these inherently damaging processes. Others—those prone to accidental death—would be evolutionarily foolish to waste energy on elaborate long-term defenses. They should get on with immediate reproduction and damn the cost. Shortly, I will discuss what we know about these defense systems, and how we might be able to make use of them to eventually slow the aging process in humans.

If reduced vulnerability to environmental dangers is in fact the explanation for slower aging in birds and bats, other groups of animals with limited environmental vulnerability should also be exceptionally long-lived. Turtles, as noted, are the longest-lived vertebrates. A particularly safe situation would seem to be living at the ocean bottom, where the climate changes little, if at all. Having a hard shell like a turtle's would seem to make one virtually impregnable to environmental accidents. I have just described the quahog, a thick-shelled, bottom-dwelling clam that has the longest life yet recorded among animals—more than 200 years has been reliably documented.

What about some specially protected groups of mammals that we haven't yet thought about? Consider mammals that may be halfway to flight—gliding mammals such as flying squirrels. Flying squirrels are only one of several groups of animals that share a similar lifestyle. Every continent except Antarctica has gliding mammals. In addition to several groups of squirrels, there are at least three groups of gliding marsupials and a small obscure mammal group called colugos, or flying lemurs, which are actually neither lemurs nor capable of flight. In addition to sharing the gliding habit, all these species are nocturnally active and live in hole nests high in trees. There is a lot of reason to suspect, then, that they will be relatively less vulnerable to predators and perhaps other environmental hazards than their terrestrial relatives. As you might now have guessed, in captivity all gliding mammals live substantially longer (and reproduce substantially more slowly) than their body size by itself would suggest, although they don't live as long as bats or birds.

To give some idea of how long gliders live, my wife, a veterinarian, once had a geriatric flying squirrel brought to her for a variety of ailments. It was aged, without question. It had cataracts and kidney and liver problems, and it wasn't long for this world. On the other hand, it was 17 years old. Flying squirrels are a bit bigger than mice, not as large as rats. Recall that mice and rats live two to three years. Even other rodents more closely related to flying squirrels, such as chipmunks, live only seven to eight years.

What about other species with special protection against predators? Animals with protective spines, for instance? They are exceptionally long-lived in captivity, as well. The egg-laying, short-nosed echidna of Australia is about the same size as an opossum and curls into an impregnable bristling ball when threatened. It is one of the very few mammals to live in captivity for more than 50 years, and it is less than one-tenth the body size of any other mammal so long-lived. Porcupines are the longest-lived rodents, most species surviving more than 20 years in captivity. Indeed, all the spined mammals are exceptionally long-lived.

If evolutionary aging theory can point out specific animals that are longer-lived than otherwise expected, can it do the same for animals that might be shorter-lived than expected? One obvious place to look is among birds with limited or no flight capability. If the association between flight and slowed aging is due to the safety that flight confers, then flightless or even weakly flying birds should be shorter-lived than your standard-issue songbird. Indeed, this is true. The shortest-lived bird, the common quail, is about the same size as a flying squirrel. It lives up to seven or eight years, shorter than the life span of a flying squirrel but still longer than average mammals its size. It is a very weak flyer. In fact, other weak flyers related to quail—chickens, grouse, turkeys, and pheasants—are similarly short-lived. We know relatively little about the lives of completely flightless birds. The 200-pound ostrich has survived only 50 years in captivity, compared with the much longer lives of smaller ravens and parrots. Penguins, which weigh three to 30 pounds, live only about as long as many songbirds.

Evolutionary aging theory explains a great deal about the general patterns of aging we see in nature. It makes clear that even though some processes inherent to life are fundamentally damaging, given the proper ecological circumstances, natural selection would have designed defenses and repair abilities to combat that destruction. We should be able to predict which animals will have exceptionally effective defenses and which will have minimal ones. Can we use this knowledge to help us find ways to slow our own aging? We take that up in the next chapter.

8

What Processes Cause Aging?

*Life's three universal rules—(1) You can't win. (2)
You have to lose. (3) There's only one way to get out
of the game.*

AN ANONYMOUS CYNIC

E ddie Andrade was a fitness nut's fitness nut. He boxed, jogged, bicycled, pumped iron, swam, and was a fitness coach for others. He neither smoked nor drank alcohol and followed his Pritikin diet with monastic discipline. In his daily journal, he kept track of how he was doing. What was his pulse? His cholesterol? Triglycerides? What had he eaten? How much was he lifting? How far and fast had he run that day? At 65, Andrade looked like he was constructed from steel cord. His friends thought he looked as good as he had at 30. Then just after sundown on May 18, 1993, Eddie Andrade stepped to the edge of a cliff overlooking the Pacific Ocean near a jogging trail where he had run some of his toughest miles and put a bullet in his brain. The note he left said simply, "Too much pain."[62]

The pain Andrade felt had nothing to do with an incurable or debilitating illness. He was remarkably fit for someone his age, but in the few months before his suicide his body had begun to fail, according to his own maniacally rigorous standards. His pulse rate had risen a bit. Several severe back spasms had left him writhing on the floor of the weight room. He had developed a hernia, which required minor surgery, and had become irrationally worried about prostate cancer, in spite of several negative tests. Andrade had discovered that even the best exercise and diet plan could not overcome the inevitability of aging, and because he was such a perfectionist, that realization was the pain that killed him.

Perhaps if Andrade had studied astronomy he would have understood that how well his body was doing was pretty unimportant by cosmic standards. If he had known more about biology, he would have realized that what was happening to his body was as natural and inevitable as birth or the succession of seasons. It was not due to the sins of sloth and gluttony but was an inevitable product of biological evolution and of the normal process of living. Life, no matter how it is lived, is damaging to our health.

But some sorts of damage are specific to only some kinds of animals, whereas others happen to just about all animals. A type of damage that is almost exclusive to humans is the common development of heart problems with aging. Pigs also get bad hearts as they age, and so do certain types of pigeons. Horses tend to die of twisted intestines, something that almost never happens to humans. Dogs and men are prone to prostate cancer, which virtually never occurs in mice or rats. Humans and monkeys seem to specialize in Alzheimer's disease. Thinking back to the evolutionary theory of aging, such relatively specialized problems should not be surprising. If aging is at least partially due to the accumulation of damaging mutations with late-life effects, and if mutations occur at random throughout the genome, then some species—indeed, some families within any species—will be prone to developing only particular sorts of problems, and other species or families will be prone to others.

But there should also be more general and widespread processes uniting the way in which humans, hyenas, and hyraxes age. If aging is partially due to the effects of genes that benefit us early in life but then cause problems later on, and if there are a limited number of physical ways in which what is beneficial in the short-term can become detrimental in the long run, then it is these genes and the processes they control that should be general, not to say universal, in aging. Because there is little more universal in the natural world than the use of oxygen to break apart chemical bonds for useful energy, a process called respiration, this would seem to be a good place to begin looking for general processes of aging.

The energy that is released by respiration was originally provided by the sun. Plants can store the sun's energy in carbohydrate chemical bonds, a process called photosynthesis, which can be expressed in a simple chemical formula:

Carbon Dioxide + Water + Energy → Glucose + Oxygen

Glucose is a simple and phenomenally important carbohydrate for our bodies. Complex carbohydrates are broken down into glucose as they are digested, and carbohydrates are stored in our bodies as long chains of individual glucose molecules attached to one another. Before being used by our cells, these long chains (called glycogen) must once again be disassembled into glucose. Respiration is the chemical reversing of photosynthesis, and so it can be expressed as:

Glucose + Oxygen → Energy + Carbon Dioxide + Water

Now it becomes apparent why respiration is more precisely called aerobic respiration: It requires oxygen. The benefit of aerobic respiration—or, as we also call it, metabolism—is quite simply life itself. We are now beginning to understand, however, that both of the ingredients of metabolism, glucose and oxygen, lead to the formation of unavoidable and injurious byproducts that seem to be centrally involved in the aging process. The two molecules we most desperately need to live, in the end, poison us. Of course, they poison other animals, such as mice and cats, more quickly than they do us. And therein lies a tale that, if we understood it more thoroughly, might point the way to slowing our own aging and living longer, healthier lives.

Rusting: More Than Just a Problem for the Tin Man

A molecule of oxygen is inspired. It rushes in through the nose or mouth, down the trachea, and through the bronchi and bronchioles to the smallest pockets of the lungs, where it is absorbed into the blood and transported to the tissues riding aboard hemoglobin. Whether it leaps off the hemoglobin in the liver, brain, or muscles, in all cases it diffuses across cells' membranes into their interior. Within the cell, most—about 90 percent—of the oxygen is consumed; that is, it is chemically transformed within small, sausage-shaped mitochondria.

Mitochondria reside inside almost all animal cells and are the site of the vast bulk of a cell's energy production. They look, under the microscope, a lot like bacteria, which is no accident because they once *were* free-living bacteria. A billion-odd years ago, they invaded larger cells in what turned out to be the beginning of a beautifully symbiotic friendship. Because of their bacterial origin, mitochondria still have their own genes (small loops of DNA), separate from what we think of as our "regular" genes in the cell nucleus. Unlike "regular" genes, mitochondrial genes are inherited from mothers only, because sperm mitochondria do not penetrate the egg at fertilization. Animal cells each contain anywhere from a few hundred to more than 1,000 mitochondria, each of which has five to 10 identical copies of the same maternally inherited genes.

Well-functioning mitochondria are, not surprisingly, crucial to a normal healthy existence. Cyanide, the powerful poison known by its bitter-almond aroma and used in quick altruistic suicides (as well as quick nonaltruistic murders) in dozens of spy novels over the years kills precisely by disrupting the energy-producing chemistry inside mitochondria. Even minor changes to mitochondria can be serious. My veterinarian wife, Veronika, once had a puzzling case of a Labrador retriever that appeared to be in perfect health, except that after

126

30 minutes of its favorite activity, chasing Frisbees, it would suddenly collapse. There was no gradual slowing down before the collapse. One moment the dog would be running full tilt; the next it was down. You could set your watch by it. Thirty minutes on the nose. The dog was still awake and alert, it just couldn't stand up or move until another 30 minutes had passed. As it turned out, the dog had a small inherited defect in its mitochondria. Humans occasionally inherit a similar defect.

It is in the mitochondria, during the final stages of food breakdown, that trouble begins, because a few percent of the broken-down oxygen molecules fail to be incorporated into water as they should be. Instead, because of minute imperfections of biological chemistry, oxygen is transmuted into damaging free oxygen radicals, or *oxidants*. As mentioned, these molecules specialize in ripping electrons and atoms loose from other molecules—oxidizing them—thereby making them less effective at whatever their normal cellular tasks are, and simultaneously creating *more* oxidants, which damage *other* molecules. They set off a chain reaction, in other words.

When iron is oxidized, we generally call the result rust. When bronze is oxidized, as on ancient bronze sculpture, we call the green film produced a patina. When we are oxidized, we call it aging.

Actually, aging, as we shall see, consists of more processes than just oxidation, but few researchers would argue that oxidation is not a major contributor to aging.

Metabolism isn't the only source of oxidants. They are also created by so-called ionizing radiation, such as ultraviolet radiation in sunlight or the much more destructive atomic "fallout" from nuclear weapons. Death and illness due to radiation exposure is in fact due to a massive formation of oxidants inside the body, breaking our DNA to pieces.

Some oxygen radicals are so reactive that, like armed psychopaths, they attack the first thing they encounter. Because most oxidants are produced inside mitochondria, injury to virtually all mitochondrial parts is a particular danger of oxygen use. Mitochondrial injury, not surprisingly, can compromise energy production by damaging vital genes and enzymes, and as damage accumulates in the energy-producing machinery, biochemical errors can produce oxidants ever faster in a viciously accelerating cycle.

Evidence is beginning to build that mitochondria are progressively more damaged in older animals, including humans. Older humans and animals more frequently have large chunks knocked out of their mitochondrial DNA, for instance, particularly in cells that are not replaced from birth onward, such as cells in the brain and muscles (including the heart muscle). Because these cells are incapable of being replaced, damage to their mitochondria can have permanent effects. Less obvious types of genetic damage than missing chunks of DNA also accumulate in mitochondria with age. The sum of this accumulated damage may explain in part why muscle power and endurance weaken with age,[63] and may also be partly involved in some of the more serious problems of aging, such as diabetes, Alzheimer's disease, and Parkinson's disease.

Not all gerontologists are convinced that this increasing mitochondrial damage is so central to aging, though. It has been compared to skin wrinkles, which accompany aging but don't contribute to it. With hundreds of mitochondria per cell, the undamaged ones may be able to carry out cellular functions quite nicely, they theorize. But all researchers are paying attention, and the pace of research into the effects of mitochondrial damage is furious. We will know a lot more about mitochondria's role in aging in the next few years.

Damage from oxygen radicals is not confined to the mitochondria, though. Metabolism creates less reactive radicals, too, such as hydrogen peroxide, which manage to diffuse into other parts of the cell before ripping into other molecules. We use hydrogen peroxide to cleanse small cuts and scrapes, because it is an effective antiseptic, able to kill microorganisms. It works on bacteria the way radiation sickness works on humans—killing by rampant free-radical formation.

A particularly dangerous target for slower-acting oxidants such as hydrogen peroxide are our "regular" genes in the cell nucleus. Here, where about 99 percent of all genes reside, is the figurative nerve center of the cell, regulating virtually all of the cell's processes. Damage to these nuclear genes can lead to all sorts of problems, including cancer, which fortunately requires a number of separate mutations.

Bruce Ames, a California biochemist with an avuncular air and an uncanny resemblance to the former basketball coach John Wooden, has calculated that oxygen radicals damage the DNA inside each of our cells some 10,000 times per day.[64] Multiply 10,000 bits of damage per cell per day by the 100 trillion cells in our body and we are in the realm of numbers that more frequently occupy astronomers than biologists. Practically all of the damage is repaired, of course, but not quite all, and that unrepaired damage will accumulate.

DNA, whether it occurs in our mitochondria or cellular nuclei, is not the only molecule damaged by oxygen radicals. Fats are also prone to oxidative damage, as you can tell by leaving the grease from your Thanksgiving turkey sitting around for a couple of days. Smell it, taste it. It will have turned rancid and repulsive. It will have been oxidized.

Fats are essential parts of our bodies. Membranes of cells and organs are rich in fats, and important membrane properties depend upon these fats' being undamaged. We also store *energy* specifically in fat tissue. Eye pigments are modified fats, as are hormones such as testosterone and estrogen. These fats can oxidize like turkey grease. One of the more important types of damage to fats for humans is what happens to LDL cholesterol, a major blood component involved in the development of atherosclerosis.

Atherosclerosis, the most common type of what is colloquially known as "hardening of the arteries," is due to the formation of thickened areas called plaques on the inside of arterial walls. Plaque development begins early in life. Even infants less than a year old frequently show the beginning signs of plaque formation, and it worsens progressively with age. Arterial thickening decreases the elasticity of artery walls, and therefore can increase blood pressure. It can become so severe that blood flow is impaired and tissues become starved of oxygen. If this occurs in the heart's coronary arteries (a particularly plaque-prone location), the eventual outcome can be a heart attack. Blood clots can also form on plaques, which leads to strokes, if the clot is in the brain, and other blood-clot diseases. Because it is involved in these various problems, atherosclerosis, at least in Western cultures, is involved in more deaths than any other single cause.

LDL cholesterol, the so-called bad cholesterol, is a complex amalgam of fats and protein that normally circulates in the blood, distributing fat to various parts of the body. It accumulates in the arterial wall in the earliest stages of atherosclerosis, and a key event in the progressive development of the plaque may be the oxidation (that is, alteration by oxidants) of LDL to a form that is not recognized as "self" by the immune system and therefore undergoes immune attack.[65] One of the reasons that oxidants are suspected in this process is that giving antioxidants, or chemicals that inhibit the effects of oxidants, to rabbits with an inherited disposition to atherosclerosis retards the development of plaques at this early stage. This finding is also compatible with some epidemiological evidence that greater consumption of *dietary* antioxidants in the form of fruits and vegetables is associated with a reduced risk of heart attack (more on this later).

Ironically, the formation of these damaging oxidants is accelerated dramatically by another molecule that we can't live without—iron. Iron is a key component of several mitochondrial enzymes involved in energy production; it is also involved in oxygen storage in muscles. But most of the body's iron functions as a critical part of the oxygen-transport molecule, hemoglobin. A shortage of iron causes a kind of anemia, a disease characterized by muscle weakness and fatigue. I happen to know firsthand about the effects of hemoglobin deficiency, because an unwanted legacy of my field research in the tropics is periodic bouts of malaria. Malarial parasites at one of their life stages live inside your red blood cells, dining on your hemoglobin. So even after you manage to kill all the parasites in your blood and are rid of the fever, headache, nausea that they cause—and the deafness, dizziness, and disorientation from the medication—you still can look forward to several weeks of wanting to sleep 12 hours at a stretch and thinking that the walk from your car to your office is a good day's exercise. Because your body doesn't store massive amounts of iron, hemoglobin takes a while to be replaced.

Probably a major reason that your body doesn't store much more iron than is absolutely needed is that iron helps convert some relatively innocuous oxidants such as hydrogen peroxide into much more damaging types of oxidants. In fact, if you grow cells from a rat's liver inside a test tube and bathe them in hydrogen peroxide, they will be fine if there is no iron in the brew. Alas, the liver cells themselves contain iron, so they are invariably killed. We know that iron is the culprit, because if you pre-treat the cells with an agent that binds iron so that it can't react with the hydrogen peroxide, the cells live. Bacteria are rich in iron, which is why hydrogen peroxide is effective in killing them.

An interesting sideline to the otherwise gloomy story of oxygen radicals is that because evolution takes advantage of what it can, oxidants have been turned to our benefit in at least one respect. Specific types of white blood cells called phagocytes are specialized for recognizing, engulfing, and destroying foreign materal such as bacteria, parasites, or cells infected by viruses. Phagocytes perform this assassination through concentrated attack with the same powerful oxidants produced in smaller amounts during normal metabolism. After engulfing bacteria, some types of phagocytes explode like hand grenades, spraying oxidants throughout the area. Some children have a genetic inability to manufacture oxidants inside phagocytes because they lack a crucial enzyme. Not surprisingly, these children suffer from chronic bacterial infections.[66] Even if the phagocytes don't explode, having these bags of oxidants lying around has its own problems. Oxidants can leak through the phagocytes' walls or be released en masse when the cells become aged and are destroyed by the body. Chronic inflammation due to a protracted immune-system battle with some infectious agent exposes one to a constant bombardment of oxidants from one's own phagocytes, and not coincidently contributes to a substantial fraction of the world's cancers, mainly in countries unable to afford effective sanitation.

If oxygen radicals are constantly bombarding us, damaging a host of molecules critical to the effective operation of our cells thousands of times per day, how is it possible that we blunder on for decades of apparent health and well-being? Also, hasn't the preceding description of oxidant production simply validated the rate-of-living theory? If normal metabolism produces oxidants and oxidants are damaging, then why isn't aging and longevity a simple function of metabolic rate?

Part of the answer is that animals have evolved an armamentarium of defenses against oxygen radicals and the damage they cause. In some species, as we have seen, evolution will have designed these defenses to be sophisticated, elaborate, and efficient; in others, they will be rather perfunctory. Species also differ dramatically in the effectiveness with which they prevent oxidant formation during normal metabolism in the first place. Some species' mitochondria are biochemically error-resistant; others' are error prone. The key to understanding this mechanism of aging lies in understanding how these differences are achieved.

Most individual cells manufacture their own antioxidants, which transform oxygen radicals into something less harmful. Indeed, the mere existence of these elaborate defenses may actually be the best argument for the involvement of oxidants in aging in the first place. The antioxidants we produce go by a bewildering range of unpronounceable names, such as superoxide dismutase, glutathione, glutathione peroxidase, and catalase, which make wonderful Scrabble words. (Note that the antioxidant defenses produced by our own cells *are not the same* as the antioxidants from food. Dietary antioxidants will be examined later.) But the relative importance of each of these enzymes is still unclear, as is the degree to which their effects may depend on some cooperative action among them. The problem with understanding their protective role has to do with the fact that studying their chemistry in a test tube is not the same as studying it in the soupy interior of our cells, where their effectiveness will depend on where they occur (in mitochondria? the nucleus? elsewhere?), which particular oxidants they encounter, and what other antioxidants are around at the same time. As a consequence of this chemical complexity, there exists no observable general relationship between increasing antioxidant levels and animal life span.[67] Some researchers even find lower antioxidant levels in some long-lived animals. However, a variety of other observations and experiments suggest that these sorts of defenses *are* important. For instance, fruit flies and yeast cells missing the genes that produce one or more of these antioxidants are shorter-lived than those that have the genes. Also, if you put a rat in a chamber filled with pure oxygen, it will be dead in about three days from lung damage. But put the same rat in a chamber with normal air (21 percent oxygen) and gradually increase the oxygen level up to 100 percent, and the rat will live quite nicely for some time. Examine this second rat's lungs and you will find increased levels of all the previously mentioned antioxidants.

Extended life in some species does seem to be associated with increased antioxidant production. Some (but not all) long-lived strains of fruit flies and minute laboratory worms have high antioxidant levels, and a fruit fly that was genetically engineered to produce additional amounts of two antioxidants outlived genetically unaltered flies.[68]

The evidence for mammals is much shakier, though. As I said, there is no correlation between species life span and antioxidant levels. Also, restricting the calories eaten by laboratory rodents is well known to extend life and reduce disease, yet these rodents do not exhibit enhanced antioxidant levels.[69] On the other hand, these dieting rodents are found to have less oxidative damage, because by some means they manage to produce fewer oxidants for a given rate of oxygen consumption than do normally fed rodents. It seems clear that by focusing only on antioxidant levels and their effects and ignoring oxidant production itself, we are looking at only half the picture. However, although we understand the basic chemistry of antioxidants, we don't understand at all how some animals process lots of oxygen without producing many oxidants, whereas other species produce oxidants rampantly. As the foregoing evidence indicates, understanding oxidant production better would certainly lead to a better understanding of the aging process. It is an area of aging research that is underdeveloped due to the current obsession with antioxidants.

Another factor to be taken into account is that antioxidants are not the only means by which we defend ourselves against oxidant damage. There are also processes by which oxidant-damaged cells or their components are repaired, replaced, or destroyed. For instance, there are enzymes that dissolve oxidized proteins inside cells. Cells can then produce more undamaged protein. Other enzyme systems are specialized for recognizing damage to DNA, cutting the damaged parts out as if with scissors and replacing them with new intact parts. This system can work only because DNA consists of two complementary strands. If one strand is damaged, the other can be used as a template so that the correct repair can be made. However, if a cell divides before the damage can be repaired, the new cell will contain only damaged DNA, and repair will be impossible. There are also "suicide" genes inside each cell, which cause the cell to self-destruct if its DNA is damaged beyond repair.

Self-destruction and replacement is a fine mechanism for assuring that damage does not go too far, but not all cells or molecules can be replaced. Nerve cells in the brain, for instance, are not replaced after birth. Neither are heart cells or muscle cells. Molecules inside our eye lens are also very long-lived, and oxidant damage to these molecules ultimately leads to cataracts. Tendons and ligaments are composed mostly of collagen fibers, molecules that are not generally repairable or replaceable. That is why athletes who tear a knee's anterior cruciate ligament must have it surgically repaired or replaced. The body will not repair it on its own. So oxidant damage to these types of long-lived, permanent cells and molecules is particularly dangerous.

What does this suggest about exercise? Exercise increases oxygen consumption, or, in other words, the metabolic rate. That is why exercising is so useful for losing weight. If increased metabolism leads to increased oxidant production, then should we worry about whether exercise, for all its vaunted healthful effects, also accelerates aging by increasing oxidative damage? Or are damage-control systems enhanced by exercise, therefore offsetting any potentially harmful effects? This question will be addressed later when we consider possible ways to slow the aging process.

Glucose Damage: Browning like Toast

My college housemate's mother worked as a butcher at a local market. Hers was a wonderful job—for us—because she brought us fine steaks night after night. The meat needed to be disposed of anyway, because it was too old to be legally sold. It tasted fine just the same.

I remember quite well that this meat always had a rather peculiar appearance. Before cooking, it was faintly brown on top as if it had been lightly seared. Meat changes color as it is cooked, of course, becoming golden and then brown, and I now know that this old stored meat had undergone very slowly some of the same chemical reactions that meat undergoes rapidly during cooking. And as surprising as it may seem, we undergo some of these same chemical changes as we age. We brown.

Meat browns and we brown because of glucose, the other major molecule of aerobic metabolism. This peculiar phenomenon was discovered in 1912 by Louis Maillard, a French chemist, who observed that a mixture of glucose and protein components, when heated, would turn from clear to yellow to a deep brown. This was subsequently called the Maillard Reaction and has ever since been of great interest to food chemists concerned with making food as tasty and tempting in appearance as possible. The Maillard Reaction is just the chemical attachment of glucose to proteins at places it doesn't normally belong, which makes a yellow or brown product. Glucose is sticky stuff, so it attaches spontaneously, given the appropriate temperature and amount of time.

Until recently, however, no one understood that this same process could occur at body temperature. But in the 1970s, it was noticed that people who have high levels of blood glucose due to poorly controlled diabetes also have glucose attached to some of their hemoglobin, a protein. That is, their hemoglobin is modified as if by a Maillard Reaction. Doctors had noticed for years that uncontrolled diabetics seemed to undergo something resembling accelerated aging. Many of the common ailments of aging, such as cataracts, atherosclerosis, heart attacks, strokes, lung and joint stiffening, appeared earlier in diabetics. Anthony Cerami, then a biochemist at Rockefeller University, put these observations on diabetics together with the Maillard Reaction and concluded that aging itself might be partially due to Maillard, or browning, products accumulating at a slow rate in the body.[70] What's more, he noted that the end result of this reaction was a series of unalterable new chemical structures in our tissues, which he called AGEs, a clever acronym from his more opaque chemical term, Advanced Glycosylation End-products.

But why would simply attaching glucose to proteins at places it doesn't belong lead to the sort of deterioration we call aging? The first reason is that some proteins form the structure and support of our bodies. Many of the most important *structural* proteins, such as my opossums' collagen, live for decades in our bodies without being broken down and replaced, without their molecules turning over. An apparently general characteristic of aging is that many of the most dramatic deteriorations of aging take place in exactly those tissues that are stuffed with long-lived, nonrenewing cells and molecules.

So let's think about collagen again. Remember that it is a flexible protein composed of three strands coiled around one another like a cable. Its flexibility is what makes it so useful for forming the foundation of arteries, veins, lungs, and skin; for making tendons and ligaments that twist and bend without breaking; for forming cartilage that cushions our joints with its resiliency. But as glucose attaches to collagen, it forms bridges or crosslinks between strands of a single collagen molecule and between molecules. As these bridges multiply over time, collagen's flexibility gradually disappears. It turns yellow and stiff and no longer makes such wonderful lungs, tendons, ligaments, or support for artery walls. What's more, collagen with attached glucose in the walls of arteries acts like an opened-jawed bear trap. It seizes and holds on to passing proteins. In this fashion, browning may play a part in the trapping and accumulation of LDL cholesterol in the artery walls—an early stage of atherosclerosis.

Proteins do many other things in the body besides provide its structure. They turn genes on and off, direct cell replication, and chaperone other molecules to their appropriate site of action. As enzymes, they are essential for virtually all of the chemical activity of a cell. The fidelity with which proteins carry out the functions they were designed for depends on their being chemically unaltered. When sugars attach to proteins inappropriately, they can impair their function and therefore disrupt the proper working of the cell. Attached sugars also make proteins less soluble in the body—more likely to solidify and become nonfunctional, and less likely to be broken down by chemicals designed to destroy damaged molecules.

Solidified proteins glued together into large masses, as it turns out, compose the characteristic brain lesions of Alzheimer's disease, which will arguably become the most serious problem of aging over the next several decades. Alzheimer's disease and other mental debilitations increase dramatically with age, particularly after age 50. The risk of mental debilitation doubles about every five years in late life—faster than the overall mortality-doubling rate, which is seven to 10 years. By age 85, as many as half of us will be mentally impaired to some degree. Therefore, as populations around the world live longer over the next decades, the number of people who will be unable to take care of themselves, unable to recognize their spouses or children, and unable to control their own bodily functions will increase at a depressing rate unless new medical treatments are developed.

The damaging brain lesions of Alzheimer's disease—so-called tangles and plaques—consist of proteins that are common and normal in the brain and that become damaging only when they solidify and aggregate into these gluey masses. Recently, browning products have been found in both tangles and plaques, and it may be that browning products themselves are involved in development of the plaques and tangles.[71]

The impairment of proteins isn't the only potential problem associated with glucose. Glucose can also bind directly to DNA, although it does so more slowly than it binds to proteins. Nevertheless, in nondividing cells, such as those composing much of the brain and heart, DNA is a long-lived molecule on which AGEs can potentially accumulate. In principle, AGE attachment to DNA could disrupt the production of new cellular proteins, could interfere with DNA repair, and could even cause mutations. As of now, however, relatively little is known about these particular processes.

Chemical theorists tell us that the Maillard Reaction should proceed at a rate pretty much determined by the concentration of sugars and proteins, and by the temperature at which these ingredients are kept. To the extent that browning is a central process of aging, the antiaging impact of caloric restriction might be partially explained by a reduced browning rate. If you feed laboratory rats only 60 percent of the calories they would eat if given unlimited food supplies, their blood glucose is reduced by about 25 percent, their body temperature declines by several degrees, and their aging is retarded by about 20 to 25 percent.[72]

But the larger picture is much more complex. Most mammals maintain about the same concentration of sugars and proteins in their bodies and live near the same 98.6° that humans do, and yet some mammals accumulate browning products rapidly and live only a year or two, while others brown slowly and live many decades.[73] How can this be?

As was the case with oxidants, the production rate of browning products is apparently only part of the story. If they accumulate at different rates in different species, then there must exist protective mechanisms—anti-AGEing processes—that are well developed in some species and poorly developed in others. Discovering the nature of these anti-AGEing processes, and being able to modify them pharmaceutically, should hold great promise for understanding and modifying the aging rate itself.

Currently, we know very little about these processes. We know that certain natural plant compounds, as well as synthetic drugs, can inhibit the formation of browning products in a test tube.[74] So it seems likely that our bodies will also contain an array of anti-browning chemicals, although they are as yet unknown. But at least one of the body's phagocytes, or scavenger cells, seems specialized to devour proteins or cells that have been "browned."

If oxidation and browning are two important general processes of aging, it might come as no surprise that they seem to operate cooperatively to our detriment. The damage caused by one affects the other.[75] So glucose and its derivative products can react with other chemicals to produce free radicals, and free radicals can accelerate browning. Also, glucose can attach to cellular antioxidant enzymes and, by doing so, inactivate them, leading to higher levels of free radicals and the damage they cause.

These two processes are likely to be centrally involved in aging, but there is no reason to expect that they completely explain it. Evolutionary theory suggests that there will be many mechanisms of aging. Browning and rusting may just be among the most general and easiest to identify. One other general process that has short-term benefits but long-term hazards is the continuing ability of cells to divide throughout life. The failure of proper regulation of this process leads, of course, to cancer, which turns out to be perhaps the most general disease of aging in the animal kingdom.

Undone by Repair Processes: The Ubiquity of Cancer

"But sharks don't get cancer," a friend of mine informed me several years ago as we were discussing the generality of diseases in the animal kingdom. He had learned this fact in a fish biology course that he was taking from an eminent professor at a well-known Ivy League university. His professor was a world expert, and knowing as little about fish as I do about extraterrestrial life, I accepted the news, but with some surprise. Because a number of my colleagues were involved in cancer research and I thought this bit of shark lore might interest them, I asked my friend whether he could get his hands on a scientific reference to this fact. After several weeks of scavenging the library for information on cancer resistance in sharks, my friend somewhat sheepishly approached me with a list of at least a dozen cancers that sharks *were* known to get. His professor, the eminent fish biologist, had been hooked by a fish story.

One reason that sharks, like many other animals that live more than a few weeks, *do* get cancer is that many of their cells have to retain their ability to divide throughout life. Cells such as skin cells, or cells lining the digestive tract, are scraped off or worn away and need to be replaced. The healing of wounds requires cell division so that new cells can replace the injured tissue. Also, because blood cells are dying all the time, sharks and humans need to produce new ones. White blood cells in particular must be able to divide rapidly on demand in order to protect us from invaders such as viruses and bacteria. So continuing cell division throughout life is crucial to many body functions.

The disadvantage of cell division is that any cell which retains the capacity to divide at the same time retains the capability for uncontrolled division. Well-regulated cell division is like the controlled operation of a motor vehicle. And just as any moving vehicle can become a lethal weapon given the appropriate malfunction, such as a sticking accelerator or failing brakes, any dividing cell can become cancerous if something goes wrong. The more times cells divide, like the more miles a car is driven, the greater the likelihood of a malfunction. Not surprisingly, you can run down the list of the most common cancers and relate them to those cells which most consistently divide—skin, digestive system (colon and rectum), reproductive system (breast, prostate, uterus, ovary), and immune system (leukemia, lymphoma).

You can always leave your car turned off in the garage, though. And some cells are analogous to turned-off cars. Neurons, brain cells that convey our thoughts and feelings, lose their capacity to divide by birth—at least in mammals. Heart-muscle cells do the same thing. Their chances of dividing out of control are consequently minuscule. Virtually all brain cancers develop in childhood soon after the cell division necessary for brain development is completed, or else they appear in supporting cells other than neurons. Heart tumors are also very rare. So one price to be paid for a body that can repair itself is the danger of cancer. Because some animals survive millions of times more cell division than others without developing cancer, however, evolution must have invented ways of effectively preventing unchecked cell division.

What Does This Mean for Medical Research on Aging?

Now that some processes central to aging have been identified, research can begin on actually altering the aging rate itself by tinkering with these processes. One of the questions we need to consider, though, is whether certain animals will be more useful than others for helping us learn how to slow aging down. This question involves the larger question of the relevance of studies of animal health and disease for humans in the first place. Some who would like to stop the use of animals in medical research would claim that there is no relevance. This is clearly untrue. Few modern medical advances did not originate in studies of animals. On the other hand, some research-ers pretend that there is no medically significant difference between a rat, a mouse, a cat, and a human. This is also clearly fallacious. The truth, as usual, lies somewhere in between. To assess precisely where the truth may lie, let's consider the development of cancer and what we know about it in mice and men.

As far as we now understand it, cancer develops in both mice and humans from a single genetically mutated cell, which undergoes a series of additional mutations as it divides repeatedly. A substantial fraction—on the order of one-fifth to one-third—of both mice and humans die from cancer, if they are lucky enough to live in safe environments such as medical laboratories or modern industrialized societies that protect them from the vagaries of nature. But let's look at how different the risks for cancer actually are in the two species.

Medical textbooks erroneously claim that the greatest risk factor for developing cancer is age. Age is indeed a major factor. At age 80, the chance of dying from cancer is about 200 times as great as at age 20. However, as Richard Miller, a University of Michigan immunolo-gist who is one of America's most trenchant and witty gerontolo-gists, likes to point out, the greatest risk factor for cancer is your species. If your parents were mice, each of your cells is some 100,000 times more likely to turn cancerous than it would be if your parents were human.

How did he arrive at this conclusion? First, mouse and human cells, with few exceptions, are about the same size. Because humans weigh about 3,300 times as much as a mouse, they contain about 3,300 times as many cells. So if the right mutations in any one of those cells can transform it into a lethal cancer cell, and if, like mice, one-fifth to one-third of humans died of cancer in two to three years, we could confidently claim that because we have so many more cells that *could* have become cancerous, each of our cells must be some 3,300 times more resistant to becoming cancerous as a mouse cell. But we *don't* develop cancer to the same extent as mice for some *30 times longer* (60 to 90 years versus two to three years)! Therefore, each of our cells must be 3,300 × 30, or about 100,000, times as re-sistant to turning cancerous as each mouse cell!

This observation is perfectly consistent with what happens when we grow mouse or human cells in a laboratory dish. Normal mouse cells divide normally for a time but then spontaneously transform into cancer cells, after which time they continue dividing indefinitely. Unlimited growth and division are cancer cells' specialty. Normal human cells rarely, possibly never, make such a spontaneous transformation in a laboratory dish. Human cells will divide for a time— a long time—but then forever cease dividing. Cells taken from human cancers, on the other hand, behave like mouse cells, dividing forever like the cells from poor Henrietta Lacks.

So there are certain similarities between mice and humans, between mouse cells and human cells, but human cells possess superior mechanisms for avoiding becoming cancerous. We might be able, indeed we *have* been able, to learn a great deal from studying cancer development in mice, but mice cannot illustrate for us how we might make our own cells even more resistant to cancer. For that we need some blazing general insight into the process of carcinogenesis, or, alternatively, we need to study elephant or whale cells.

I've elicited snorts of disbelief at scientific meetings by making such a suggestion. What? Study something besides mice and rats? We know so much about the genetics of mice and the physiology of rats. We've got millions of animals sitting on the shelf available for research right now. What could the study of elephant or whale cells possibly tell us?

What indeed? Elephants contain about 40 times the numbers of cells we do, and whales as many as 600 times as many cells. Yet elephants and whales live, to a reasonable approximation, just as long as we do. Therefore, their cells must be 40 to 600 times *more* resistant to turning cancerous than our own. Could we perhaps learn something about cancer resistance from studying these cells?

This question goes to the heart of the logic about why we use animals in medical research on specific problems or diseases in the first place. One reason, of course, is that we (most of us, anyway) are willing to do things to animals that would be ethically objectionable to do to humans. For one thing, we can raise 1,000 genetically identical mice (rendered genetically identical by many generations of brother-sister matings) under identical conditions, completely controlling their access to food, water, mates, and social interaction, while giving half of them, say, a new drug or a suspected carcinogen. In two to three years, we can track them through their entire lives, observing how the development and outcome of disease differs between groups. We use mice and rats because they are conveniently available and because decades of medical research has taught us more about them than we know about any other species, with the possible exception of ourselves.

But rats and mice are only two of more than 4,000 species of mammals; moreover, they are closely related to each other and not particularly closely related to humans. Studying only these species to learn all about our medical problems is analogous to trying to learn all about the psychology of American corporate CEOs by interviewing the same pair of Tasmanian sisters again and again. Yes, if you are a clever interviewer you will virtually always be able to learn new things, but at some point you will learn more by interviewing someone else.

With respect to learning about aging, we know that mice and rats are relative failures at combating the major destructive processes of life. That is, they only live two to three years because they are so inept at combating oxidative damage, browning damage, and all other problems of aging. We can learn, and have learned, a great deal from them, and we will continue to do so. But might we now profit by interviewing someone else, perhaps a species that is *better* than we are at dealing with the damaging processes of aging? Evolution has produced better solutions than our own. Might a peek at the nature of those solutions speed our search for how to improve our own defenses at combating life's destructive processes?

The logic here is simple. John Harrison, eighteenth-century clockmaker extraordinaire, worked for more than 30 years to solve the greatest technological challenge of his time—producing a timepiece of such accuracy as to allow sailing ships to compute their longitude at any location in the world.[76] Harrison had to proceed from his knowledge of inferior clocks, deducing little by little how they might be improved. Might he not have produced a superior clock much more quickly if he had stumbled upon an existing timepiece of superior accuracy, perhaps left behind by Martian astronauts, so that he could have tinkered with it and studied its design and learned from it?

It obviously isn't very practical to have a laboratory colony of whales or elephants. Yet for a lot of kinds of research, you really don't need a laboratory colony. Small-skin biopsies like those taken from humans hundreds of times per day in doctors' offices around the world would suffice to provide whale or elephant cells to grow in the laboratory. Zoo and marine-park veterinarians take blood and biopsies all the time for routine medical surveillance and treatment. There is no reason why bits of extra tissue couldn't be taken to help us understand cancer resistance.

But, in fact, we could support laboratory colonies of some sorts of animals with proven superiority with respect to the central damaging processes of aging, too. Think, for instance, about birds and their long lives. Remember that birds are dramatically longer-lived than mammals. During those long lives, they consume as much as five times as much oxygen as humans. How do they avoid being overwhelmed by oxidative damage? They have blood sugar levels that are two to 10 times higher than those of mammals and live at a body temperature 6° to 7° higher. How do they avoid having their tissues look like overdone toast as they move into late life?

There are few bankers who are as conservative as the medical research establishment. That establishment, however, has been wildly, and increasingly, successful over the past 50 years at gaining an understanding of specific diseases and how to treat them. So telling the establishment that aging might profitably be approached in some dramatically new way is a tough sell. On the other hand, it is arguable whether the medical establishment has had much success at understanding or treating aging during the same period. The major theoretical advances in understanding aging took place in the 1950s and 1960s. The one uncontestable method for retarding aging, and extending life, in laboratory rodents—restricting the number of calories that they eat—was identified early in this century, and we still don't understand how it works. We also still do not have antiaging pills or specific therapies that might slow the aging process. Maybe additional species do have something to teach us, after all.

Everyone will admit that there is at least one aspect of aging about which rats and mice can tell us virtually nothing. I've mentioned it in passing several times, but with our focus on longevity and survival so far, it is easy to forget that some aspects of aging have little to do with staying alive. Nevertheless, aging in our reproductive system is likely even more crucial to the way we organize our lives than the increasing likelihood of dying. Reproductive aging happens to us all, but it happens most dramatically to women. Half of the human species lives more than a third of its life having lost the ability to reproduce. Who knows what far-reaching social manifestations this simple fact has? Rats and mice do not have menopause; humans do. Can some knowledge about evolutionary trade-offs and general processes of aging help us understand menopause and its health consequences? Let's see.

9

Reproductive Aging, Menopause, and Health

Deep in the lush Kibale Forest of western Uganda in early 1973, Peter Waser, a doctoral candidate at Rockefeller University, quietly watched an elderly female monkey whom he called Kink-tail resting by herself as her troop raced about in the trees, plucking figs and eating them. Waser wasn't sure of Kink-tail's exact age, but there was no doubt she had led a long, hard life. Her name came from her distinctive tail, which had been broken in at least three places. Her barks were hoarser and lower-pitched than those of the other females; her nipples were long, slack, and gray; her fur was scruffy; and she was scarred in many places.

What made Kink-tail, a type of monkey called a gray-cheeked mangabey, especially interesting to Waser was that she had apparently undergone menopause some three years earlier. Field biologists almost never encounter animals that have outlived their physical capacity to reproduce, which makes good evolutionary sense. If animals *did* routinely live much longer than they were capable of reproducing, any new mutations that extended reproductive life would be at a huge advantage and should soon spread and become ubiquitous, and so the period of reproduction would become longer until it matched the length of time that animals could expect to live.

But Kink-tail, despite this logic, *had* apparently outlived her reproductive system. During the several years Waser had been watching this troop, the skin around her vagina had never grown red and swollen as it did in the other females as they became fertile. Infants or juveniles had never associated with her in that special way they do with their mothers. Also, the males in the group showed no sexual interest in her, even on the few occasions when she solicited their sexual attention. Kink-tail was beyond the age of menopause. Waser left Uganda that April, and when he returned 15 months later, Kink-tail had disappeared and was never seen again.

Because finding a postmenopausal monkey in the wild was so unusual and unexpected, Waser reported his observations of Kink-tail in the scientific literature[77] and queried his friends who had spent years watching primates. Had they ever seen postmenopausal females? Jane Goodall said, yes, in Tanzania's Gombe National Park, she had observed several female baboons who seemed to have gone through menopause. And Flo, the oldest female chimpanzee she had observed, stopped cycling about two years before she died. Sarah Hrdy, another researcher, said that she had also seen a few postmenopausal females during her study of langur monkeys in western India. On the other hand, Jeanne and Stuart Altmann, having watched baboons for several decades around Kenya's Amboseli Reserve, had never seen a postmenopausal female. It looked as if Waser's observations were fairly typical. Menopause in nature occurred, but only to the exceptionally lucky few who lived a bit longer than they were designed to. Natural menopause, in other words, was always rare enough to be noteworthy.

By contrast, of course, living past menopause is anything but rare for human females. Typical women in industrialized countries can expect to live more than one-third of their lives beyond their reproductive years. But what about humans in a primeval state of nature? Before sewage systems and supermarkets and doctors with their life-saving surgeries and antibiotics, had menopause been a routine part of life or was it also a noteworthy exception to the common course of events, as with monkeys? And if it was routine, how could such a seemingly evolutionarily disadvantageous trait be explained?

These are not necessarily abstract questions. The answers to them probably matter in some important medical ways. For instance, if living long enough to experience menopause has been a routine part of human life only during the past few centuries, then evolution will not have had time to adjust women's bodies to living with reduced postmenopausal hormone levels. That is, if over the vast sweep of evolutionary history, women almost always experienced only the high estrogen and progesterone levels of the reproductive years, then it is likely that their bodies would be designed to operate best in such a hormonal environment. Accordingly, medically replacing the hormones lost at menopause might be expected to help extend life and alleviate a range of major and minor health and welfare problems of postreproductive life. It's true that after menopause women still die at only about half the rate of similar-aged men, but maybe they could be doing even better.

On the other hand, if our evolutionary history differed substantially from that of monkeys and apes—if we have lived with menopause for countless thousands of generations because it has some special adaptive value—then women's bodies are likely to be physiologically adapted to the altered hormone levels of menopause in many subtle ways. Fiddling with that physiological adaptation would be a dangerous game, perhaps upsetting some delicate internal balance, conceivably leading to unexpected health-damaging consequences.

But why is the focus on females only? Is this really an issue just for women? What about the decline in male hormones with aging? Can we really consider the sexes different with respect to reproductive aging?

Is There Really a Sexual Difference in Reproductive Aging?

If women die at lower rates than men from the time of conception into old age, does it really make any sense to think that they might age more quickly than men in one respect—reproduction?

148

The answer is yes. The reproductive system of humans deteriorates with aging, as do all other bodily systems. For men, reproductive aging follows the same pattern of gradual deterioration as sprinting speed and language-learning ability. For instance, the rate of impotence gradually increases after age 30, so that by age 70 impotence strikes almost half of men. The other side of this statistic, though, is that more than half of 70-year-old men can still sire children, if they have the energy and inclination.

But no 70-year-old woman is in danger of having children. For women, reproductive aging is puzzlingly different—dramatically accelerated. Although women, like men, have the potential for sexual arousal and orgasm throughout life, they lose the ability to reproduce at about age 50. And menopause itself is only the final step in a process of rapid reproductive decline that begins in women's twenties. Although more women are delaying reproduction until their thirties or forties, relative to a reproductive peak at age 25, even 35-year-old women have considerably more difficulty becoming, and remaining, pregnant. Also, the chances that a woman's baby will be genetically defective, or that the mother will die in childbirth, doubles by age 35, compared with women in their early twenties. Menopause, the halting of menstrual cycles, is only the final step down the staircase of reproductive aging for women.

The immediate reason that women stop cycling when they do is that their ovaries become depleted of eggs. Unlike many animals, humans and most other mammals stop making eggs before they are even born. So the 7 million or so eggs already contained in the ovaries of a five-month female fetus are all she will ever have. And these eggs begin dying immediately. Only 1 million to 2 million are left by the time of the fetus's birth. By puberty, when the eggs begin to be shed from the ovaries monthly so that they can potentially be fertilized, only 250,000 (or one-thirtieth of the original number) are still left. During each menstrual cycle thereafter, one egg will fully develop and be ovulated, while at the same time a host of unused others will die. Women lose eggs at a constant rate between puberty and about 35 years of age, when suddenly, for reasons we don't understand, the rate of egg loss accelerates about twofold. By menopause, few eggs remain. And because estrogen is secreted only by the cells surrounding developing eggs, the ovaries no longer produce sufficient estrogen to drive the menstrual cycle to completion. Hence, cycling and further egg development ceases. Menstruation stops. Hormones produced by the ovaries, estrogen and progesterone, plummet. And that is that.

So as much as it might be socially or politically preferable for men and women to age reproductively in a similar manner, they don't. There has been considerable anger, denial, and ideological ranting surrounding this stark fact. Biology and personal ideology sometimes conflict, and you ignore biology at your peril. And much as we might wish it, men do not have menopause or anything like it. What is sometimes called "male menopause," when an aging man may experience his first real psychological depression or develop impotence or a taste for sporty cars and younger women, is a psychological syndrome, not a physiological inevitability. Men continue producing sperm throughout life. They experience no dramatic fall in testosterone as they age. When testosterone declines, its decline is gradual, about 1 percent per year, on average. Some men may not experience a decline at all.

So although there may be no doubt that the sexes differ in their reproductive aging, what there *is* doubt about, is what this difference means about how our bodies have evolved.

Why Menopause?

There are two conflicting ideas about why women commonly undergo menopause. What I'll call the Blessings of Modern Life Theory of menopause, which is probably the dominant theory in the medical community, is that it really needs no explaining at all. This theory assumes that throughout our prehistory and until relatively recently, people could expect to live only 20 to 40 years. That is what the evidence of the bones tells us. Remember the 1,300 skeletons excavated at the Libben Site near the shores of Lake Erie, which indicated that only about one newborn in 50 lived to 50 years of age?

Other archeological digs in Africa and the Middle East indicated the same thing. And Neanderthals are assumed by many anthropologists never to have lived beyond 40 or 45 years. So women's bodies have been designed by evolution to produce about as many eggs over as long a period as they would ever need to, and menopause is just an incidental byproduct of the extra longevity of modern life—something our bodies were no better designed for than serving a tennis ball, reading for hours at a time, or typing all day on a computer. As a consequence, it's no surprise that we develop some of the medical problems of modernity, such as tennis elbow, myopia, carpal tunnel syndrome, or, in the case of postmenopausal life, hot flashes and osteoporosis.

According to this idea, humans are no different than animals kept as pets or in zoos, where they are pampered and protected and as a consequence live much longer than they would in nature. Menopause, or something like it, happens to captive animals from rats to guinea pigs to pigeons, parakeets, or emus.

The potential medical significance of the Blessings of Modern Life Theory is that if it is correct, then menopause is just like a host of other bodily changes that happen with aging, an uncontrolled deterioration of a previously well-regulated process. It wouldn't be surprising if other physical problems were associated with it. Implied by this idea is that women's bodies weren't really designed to operate on postmenopausally low levels of estrogen or progesterone, and conceivably they might operate better on premenopausal hormone levels. Thus, replacement of these hormones might possibly be a very reasonable medical intervention—like administering insulin to a diabetic or giving a movie star a face-lift.

One thing that is important to remember, though, is that there is no sudden leap in a woman's probability of dying around menopause. Like men, women of this age continue doubling their probability of dying every seven to 10 years. And they continue to die at rates about half those of similar-aged men. However, there are certain shifts in the pattern of diseases and causes of death after menopause, which we'll discuss shortly.

But not everyone is happy with the evidence of the bones. They are skeptical about our ability to distinguish a 40-year-old from a 60-

or 70-year-old skeleton, especially since the conditions of life were so different in the distant past. One reason for this skepticism is that since the early 1960s, when a small group of scientists from Harvard University journeyed to southern Africa's harsh, dry Kalahari Desert to observe the !Kung bushmen, as a way to ask questions about what human life in the primordial past was like, anthropologists have found time and again in many parts of the world that people in existing pretechnological, indigenous cultures seem to live considerably longer than the evidence of the bones suggests they should. Taking the !Kung, for instance, who were later made famous by their depiction in the movie *The Gods Must Be Crazy*, about 40 percent of !Kung women seemed to live into their mid-forties, when they typically experience menopause. They had even developed the habit of changing their child-rearing pattern as menopause approached. Whereas younger mothers typically nursed their infants only two or three years, older mothers often nursed much longer, as long as eight or 10 years, so that they would frequently be nursing right through menopause. Thus, the evidence from the !Kung and other existing cultures seems to indicate that menopause is a human universal—part of our collective experience probably since we swept out of Africa 100,000 years ago. But if the relative commonness of menopause among the !Kung and other indigenous cultures compared with other primates suggests that we have truly broken with our closest biological relatives and developed something new and novel in nature, how could something so apparently maladaptive happen? How could evolution, which favors the most reproductively successful individuals, have favored adaptations to end reproduction before the end of life? And why would it do so in only one sex?

Out of observations of indigenous people and the dramatic difference in the rate of reproductive aging between the sexes grew the second theory of menopause, which assumes it is an evolutionary adaptation specifically of women and that it acts to their ultimate advantage. I'll call it the Good Mother Theory. The logic of this idea develops from the way in which evolution will favor any trait that causes its bearers to pass along copies of their genes more effectively. The most obvious way to pass along copies of your genes is by continuing to have more children, of course, but genes can be passed along in other ways, as well. Instead of reproducing yourself, you could facilitate the spread of your genes by helping your children reproduce. Your children share half their genes with you (your grandchildren share one-quarter), so helping your children produce grandchildren could conceivably be more evolutionarily advantageous than continuing to reproduce yourself—especially if your energy could be more effectively used to produce grandchildren than more children of your own.

The Good Mother Theory attributes menopause to our long and intense requirement for child care. It recognizes that after her early twenties, a woman's chances of successfully bearing a child begin declining for a variety of reasons. Difficulties with pregnancy grow more common, so that the risk that the mother or child will die increases. It becomes more likely that the child will be genetically defective. And, of course, the odds that a woman will survive until her child is independent declines because of the increasing chance of dying as we age. Remember, we are not considering the magnitude of these risks today when transfusions, antibiotics, cesarean sections, and a host of other medical procedures have made the whole birth process relatively safe. We are talking about late-life childbirth over the premedical millennia, during which evolution designed our physiology. If just a century ago women died in childbirth at more than ten times the current rate, what might those rates have been when we lived nomadic hunter-gatherer lives?

Ultimately, so the theory runs, a woman's chances of successfully reproducing decline to the point that it becomes evolutionarily more advantageous (that is, a more effective way of passing on her genes) for her to halt her own reproduction and shift her energy to raising her last child or helping her older children raise *their* children—being, in other words, a good mother and good grandmother, too.

Whereas the Good Mother Theory may or may not apply to humans (we will assess the evidence shortly), it *is* likely to explain the few examples of routine and natural menopause among mammals—that found among certain whales. Technically, of course, whales do not have menopause (which literally means "the cessation of monthly bleeding"), because they do not have monthly bleeding, but they unquestionably have a substantial life after females' ovaries are depleted of eggs and they no longer cycle. So let's call it menopause and look at how it works in short-finned pilot whales.

Pilot whales—or blackfish, as they are also called—are small (10 to 25 feet long) toothed whales best known for their tendency to fatally strand themselves in large groups on beaches throughout the world. They may be among the most socially complex of mammals. Sleeping during the day, feeding on squid and small fish at night, maneuvering within an intricate social network of movers, shakers, peers, elders, and several generations of relatives, their mutual attentiveness may in some cases be their undoing. Swarming toward the cries of wounded or frightened animals, as they are wont to do, may actually cause their strandings, as they will rush toward any accidentally stranded animal. For whalers, this socially responsive and responsible habit has meant that a single harpooned animal dragged into shallow water will draw in many others, where they can be easily dispatched.

Ironically, these mass strandings and large fishing (or more correctly "mammaling") kills, combined with our ability to tell how old animals are from treelike growth rings in their teeth, have provided us with virtually everything we know about pilot whales, including how their life course is eerily like a somewhat telescoped human one. For instance, females reach puberty when they are about 10 years old and have their first calf not long after. Female reproduction peaks in the late teens, and then declines rapidly. By their mid-twenties, they are reproductively aged, and by no later than age 40 their egg supply is exhausted and they stop cycling. What is amazing, when pilot whales are compared with other animals in nature, is that about a quarter of all females live to this age and those that do can expect to live another 14 years.

Also very humanlike, male pilot whales reach puberty somewhat later (mid-teens) than females and die at higher rates throughout life. However, the difference in longevity between the sexes is dramatically larger for pilot whales than for humans. At birth, females can expect to live 23 years, compared with 12 years for males. By the age that females reach menopause, when a quarter of them are still alive, only about 5 percent of males survive. The oldest male identified so far is 46 years old, compared with 63 years for the oldest female, and as far as we know there are no high-stress, type A occupations for pilot whales. Just maleness itself seems to be enough of a burden.

What do postmenopausal females do with themselves for the 14 years or so after they have finished bearing young? Like the !Kung, they seem to spend a considerable fraction of their lives looking after their last born. Young pilot whale mothers tend to nurse their calves for only five years or so, but as menopause approaches, they nurse longer. Female calves of older mothers nurse up to about eight years of age, but some males have been discovered still nursing from their mothers in their mid-teens—on the cusp of puberty—giving fresh meaning to the term "mama's boy."

Mammals will continue to produce milk only as long as they are regularly suckled from. Therefore, this long-term nursing suggests a continued very close and constant relationship between mother and son. Whether menopausal mothers continue helping their last offspring in ways other than providing milk is unknown, but it is certainly possible given the close and continuing contact between them.

Killer whales, too, seem to experience menopause and to live in a similarly complex social web. Female killer whales remain among close relatives throughout life. So cohesive are these extended family groups, called pods, that each has a recognizable vocal dialect that persists over generations.

Although both sexes of killer whales reach physical maturity in their mid-teens, the sexes differ in their life expectancies, just like pilot whales. At birth, males can expect to live about 26 years, with a few surviving to their fifties or slightly beyond. Females, at birth, can expect to live 46 years, and some apparently reach 80 to 90 years of age. When they reach menopause at about 40 years of age (when 60 to 70 percent of them are still alive!), females can expect to live another 25 years.

We know relatively little about the form that maternal care may take in killer whales. They don't appear to have nursing until puberty, as some pilot whales do. They seem to nurse no more than a couple of years. However, along the coast of Washington State and British Columbia, where killer-whale groups may contain as many as four generations of closely related females, when a new calf is born the previous calf shifts its behavior from maintaining physical proximity to its mother to a similar proximity with a postmenopausal female, most likely its grandmother. It is difficult to explain this social shift unless the grandmother is providing the same sort of care and assistance to the calf that the mother previously had, but we have no clue as to the nature of this assistance.

So although we don't know many of the details, it appears as if menopause among whales fits the Good Mother Theory. Certainly, there is nothing in whales' recent history to suggest a sudden extension of life necessary for the Blessings of Modern Life Theory to be invoked. Unlike that of humans, whale hygiene has probably not improved in recent times, and their medical practices do not seem to be overly sophisticated, so far as we know. If anything, you might expect that whales are living shorter lives than they did in prehistoric times. Pilot whales have been hunted by humans for at least 400 years, probably most successfully in recent times. But killer whales have never been hunted in enormous numbers, and are even less so now.

By default then, and by assuming that the continued association of young whales with menopausal females poses some sort of help, we can assume that these whales are acting as Good Mothers.

A medical implication of the Good Mother Theory is that far from being an uncontrolled deterioration of a once well-regulated reproductive system, menopause represents a more controlled reproductive cessation—by analogy, more like the controlled, smooth descent of a plane with failing engines than a tumbling, fiery crash. Evolution would be expected to have shaped female physiology to thrive to a certain extent with the reduced menopausal levels of estrogen and progesterone, because they could be Good Mothers, after all, only if their postmenopausal health were as good as possible. Attempting to modify hormone levels—returning them to their premenopausal state, for instance—might be trying to fix a well-operating engine.

Blessings of Modern Life or Good Mother?
The Human Evidence

What does the evidence suggest about the believability of these two theories about human menopause? First, it's unlikely that we will settle the heated issue of whether ancient humans lived dramatically shortened lives, as the bones suggest, or whether as the study of living cultures suggests living to 50 years of age has been common for millennia. But regardless of how long we lived in the past, one thing that can be addressed is whether women after menopause are capable of providing enough assistance to their existing children and grandchildren to compensate in evolutionary terms for the loss in personal reproduction that menopause entails.

One attempt to answer this question by trudging into the field and living with modern indigenous people came from the anthropologists Kim Hill and Magdalena Hurtado, who in 1980 began spending time with the Ache of the eastern Paraguayan forest.[78] The Ache were hunter-gatherers who had been relatively uninfluenced by the outside world until the 1970s. Having observed the role of older female Ache in assisting their children and grandchildren, and having collected information on which children failed to live into adulthood, Hill and Hurtado estimated the effect of this help on the survival of children and grandchildren and came to the conclusion that unless their estimates were dramatically wrong, menopause as a method of passing one's genes was decidedly inferior to reproducing until life's bitter end. Older women simply didn't provide enough aid to others to offset their own reproductive loss. Hesitant to unconditionally reject the Good Mother Theory, which to most biologists is the more interesting idea, Hill and Hurtado concluded that *if* in the past older women could have improved the survival of their children and grandchildren more than they seemed to do at present, and *if* they also improved the survival of relatives in addition to own their children and grandchildren, and *if* older premenopausal women were more likely than observed to have died in childbirth, then evolution might indeed have favored an adaptive menopause. A ringing endorsement of the Good Mother Theory this clearly wasn't.

157

A different approach was used by the anthropologist Alan Rogers of the University of Utah. Rogers is a tall, quiet, and thoughtful theorist who builds mathematical models to explore the plausibility of ideas such as the Good Mother Theory. Rogers used equations to show exactly how much help postreproductive mothers would have had to provide to their children for adaptive menopause to have evolved. The answer, it turned out, was a lot—a whole lot. If having a postmenopausal "mother" around could double the number of children that *all of her children had* and completely eliminate infant deaths, then evolution would have led to menopause by natural selection. A good mother, indeed! This would require a Great Mother— a greater mother than we currently know about in any place or at any time.

The Blessings of Modern Life Theory seems more plausible at present, if only because frank advocates of the Good Mother idea have been unable to make it seem very likely. And there is also the laboratory rat to consider. Rats show us that you don't need long-term maternal nurturing to have more rapid reproductive aging in females.

Rats are no paragons of extended juvenile dependence or parental attentiveness. Newborns nurse for only three weeks and become fully independent adults in only three months. Yet by 18 months of age, female rats have stopped cycling while more than 80 percent of males are still going strong, and some males may still be breeding a year later. In other mammals, we often find the same thing. More rapid female reproductive aging seems to be a mammal characteristic, not something specific to animals needing parental care for years.

What Does It Mean Medically?

Currently, the case for the Good Mother as an explanation of human menopause seems to be largely built upon a hope and a prayer— the hope that something sympathetic in our nature such as devotion to our children is responsible for accelerated female reproductive aging, and the prayer that evidence supporting this hope will eventually be revealed.

If, as seems increasingly likely, menopause is largely an accidental byproduct of increased modern longevity, what medical significance does this evolutionary analysis convey?

First, we can't assume that women's bodies are designed to thrive with the near absence of progesterone and a 90 percent reduction in estrogen after menopause. They probably aren't. Estrogen in particular is involved in many bodily functions besides reproduction. It affects arteries, bones, the brain, sexual behavior, sleep patterns, intestinal absorption of food, and the immune system, to name a few. All in all, about 400 different actions of estrogen have been described.[79] Therefore, it shouldn't be surprising that certain health and disease patterns besides those related to reproduction are altered after menopause.

Not all of these alterations are bad. For instance, menopause slows the rate of increase in certain gynecological cancers (breast, uterus, ovary) with age, so that prior to menopause women in the United States double their annual risk of contracting breast cancer about every three years, but after menopause it takes 13 years for the risk to double again. These odds are higher or lower for individual women, depending on their particular menopausal age and a variety of other factors, of course; but all else being equal, 45- or 50-year-olds who have already undergone menopause *are* less likely to get gynecological cancer during the remainder of their lives than are 45- or 50-year-olds who are still cycling.

Menopause affects these gynecological cancers for the simple reason that normal female reproductive hormones (estrogen and progesterone) are the underlying cause of these cancers in the first place. Consider, for instance, breast cancer, about which we know the most.

You occasionally read or hear that no one really knows the cause of breast cancer, or alternatively that only a quarter to a third of breast-cancer cases can be attributed to known risk factors. These statements are untrue. More than 99 percent of breast-cancer cases can be explained by one known risk: periodically secreting large quantities of estrogen and progesterone—that is, being an adult female. Men have breasts, after all, and if they are medically administered the appropriate hormones, men can even produce milk. But men lack the large monthly surges of estrogen and progesterone that women experience after puberty. Men, in consequence, account for only about 0.5 percent of all breast-cancer cases.

The overwhelming evidence implicating women's hormones in the development of breast cancer has been systematically compiled by the epidemiologist Malcolm Pike of the University of Southern California. Pike points out that nearly anything that increases the number of monthly menstrual cycles a woman has in her lifetime will also increase her likelihood of developing breast cancer. That means the risk increases the earlier in life cycling begins and the later in life it ends. Developing breast cancer is an unavoidable risk that accompanies the ability to reproduce. Removing the ovaries prior to menopause decreases the risk of breast cancer, and the younger the woman is when it is done, the larger the decrease.[80]

There has been a great deal of publicity surrounding the fact that breast-cancer rates have been increasing in the United States by about 1 percent per year for half a century. This trend has been attributed to nearly every aspect of modern life that someone dislikes—chemical pollution, eating too much fatty food, drinking too much alcohol, birth-control pills, electromagnetic radiation—everything but UFOs, fluoridated water, and obnoxious talk-show hosts. But despite a great deal of research, little hard evidence has surfaced to implicate any of these things. A different, perhaps more reasonable, notion is that increasing breast-cancer rates are due to otherwise beneficial aspects of modern life, such as good nutrition, the reduced necessity for back-breaking physical labor, increasing health consciousness, and the ability to manage one's own life course.

Why do I say this? Good nutrition and reduced physical labor together increase the occurrence of breast cancer by lowering the age at puberty and increasing the age at menopause. European and North American girls now undergo puberty about four years earlier than girls did just 150 years ago, and the link among nutrition, work, and the age at puberty is well established. Women require a certain amount of body fat before puberty will occur. It's no mystery that eating well and not exercising too much will increase body fat. Consistent with this is the fact that female endurance athletes reach puberty when they are older than their more sedentary classmates, as do women from impoverished countries when compared with women from the industrialized world.

Similar factors accelerate puberty in men. In New Guinea, I was told an amusing tale of two boys from a remote village who were sent away to jail for some senseless teenage crime. When they returned two years later, having eaten regularly and dawdled in their cells all that time, they were hardly recognizable. Their voices had broken, whereas the voices of their village peers had not, and they towered above even the adult men of the village. Jail had made them, quite literally, bigger, more powerful bullies than they were when they left.

Unlike puberty, there is no historical record on age at menopause, but the linkage of menopause, nutrition, and physical work is still pretty clear. We know, for instance, that women in nonindustrialized countries undergo menopause a few years earlier than do women in the industrialized west. Even in nonindustrialized countries such as Papua New Guinea, malnourished women undergo menopause earlier than do those who are better nourished. So all evidence suggests that the age at menopause has increased slightly up to the present in the industrialized world.

These same factors—nutrition and physical work (that is, exercise)—probably explain at least some of the international differences in breast-cancer rates, as well. China, for instance, has about one-quarter the incidence of breast cancer as the United States, and not coincidentally Chinese women reach puberty more than four years later than do women in the United States. Even among women who are adult and actively menstruating, there may be considerable differences in hormone levels. Hardworking indigenous Nepalese women have only about half the progesterone levels of middle-class Bostonians, even when both are cycling.[81] So compared with our own immediate ancestors and the people of many other cultures throughout the world, we are living in a hormone-sodden environment thanks to plentiful food and a life of physical ease.

Americans are also becoming increasingly obese. Obesity has a somewhat paradoxical effect on breast-cancer risk, due to its effect on hormone levels. Premenopausal obesity decreases the odds of getting breast cancer, whereas postmenopausal obesity increases them. This is because excessive body fat can suppress menstrual cycling in premenopausal women, thereby reducing hormone production. By contrast, after menopause obese women produce more hormones than do thin women, because body fat converts other hormones into estrogen.

If hormone production summed over a lifetime determines by and large one's risk of developing breast cancer, what about the effect of pregnancy? Late pregnancy is a time of very high levels of both estrogen and progesterone, yet many studies verify that the lower the age of the first completed pregnancy, the lower chance also of getting breast cancer. In fact, women who have a baby before age 20 get breast cancer less than half as frequently as women whose first baby is born after they are 30. The number of pregnancies doesn't seem to be important, only the age of the first one.

How Hormones Can Cause Cancer

In order to understand the apparent paradox that lifetime exposure to estrogen and progesterone increases the risk of breast cancer, yet early childbirth decreases the risk—and to understand how other factors such as socioeconomic class and personal reproductive choice may be implicated in breast cancer—we need to now understand *how* normal hormones increase cancer risk. Simply put, estrogen and progesterone increase the risk of breast cancer because they cause the cells lining the milk ducts in the breast to divide prolifically during the latter part of the menstrual cycle, when the body is preparing for pregnancy. When no pregnancy occurs, these newly formed cells die, returning the breast to its original condition. During the next cycle, there is another round of cell division and cell death if no pregnancy occurs.

Cell division throughout the body must be exquisitely controlled, because even small deviations from the appropriate timing or amount of division can cause problems. If the division rate for cells such as skin cells exceeds the rate of skin-cell loss by even a little, the overall equilibrium is destroyed and cell numbers will increase inappropriately. Inappropriate increases in cell numbers cause everything from warts and moles to more serious problems, such as tumors. During cell division, the DNA, or genetic material, of each cell must be copied precisely so that the original and copy will be identical as each is incorporated into a new daughter cell. Errors in the copying process are called "mutations," and all further descendants of the mutated cell will carry the same error.

But as we've already seen, the DNA within each cell is being damaged thousands of times per day by free radicals. Usually, this damage is repaired by an elaborate cellular machinery designed for this purpose. However, if a cell happens to divide before a damaged bit of DNA has been repaired, the result will be a mutated cell. Thus, the more frequently cells divide, the more likely a mutated cell will appear. The mutation may not be in a crucial segment of your DNA for controlling cell growth. It usually isn't. But on the other hand, it might.

Because controlling the division rate is so important to life, a number of partially redundant controls on the system exist—something like having a series of brakes on your car to stop you in case your accelerator sticks. Full-blown cancer requires overcoming all cellular controls on the division rate. That is, the accelerator is stuck, and all the brakes have failed. It requires, therefore, a number of independent mutations in different control mechanisms. Loss of control is gradual. If one brake fails, you can still stop, but not as quickly or precisely as before. If a second goes, you stop even less quickly. Then, if the accelerator sticks, stopping at all becomes very difficult. And if the last brake fails, you have runaway cell division— that is, cancer. This just about describes the development of colon cancer, which we understand particularly well. Colon cancer seems to require at least four mutations—one stuck accelerator (technically called an oncogene) and three failed brakes (called tumor suppressor genes).

Now it becomes apparent how a person can inherit a tendency to get certain types of cancer, and why it is only a tendency, not a sure thing. You can be born with a faulty accelerator or faulty brakes.

That is, in some families a mutated oncogene or tumor suppressor gene occurs in the reproductive cells, and so is passed along from one generation to the next. Inheriting one or more of these mutations means that you will require fewer additional mutations during your lifetime to develop cancer. So your chance of getting cancer is considerably higher than the chance for someone with the accelerator and all brakes in working order. On the other hand, mutations strike your DNA at random sites, and it may be that even with one brake missing at birth, your others won't fail or your accelerator won't stick.

Understanding cancer from this perspective, the paradox of hormone levels, pregnancy, and breast cancer can be resolved. After having giving birth once, the cells lining women's milk ducts no longer undergo those monthly cycles of rapid division followed by cell death. The cells become specialized to produce milk and remain that way. Once they are no longer regularly proliferating, the chances of additional mutations in those cells is greatly reduced, even in the presence of high levels of estrogen and progesterone. The later in life a woman has her first pregnancy, the more cycles of milk-duct cell division she has had, and the greater the chances that some critical mutations will have already occurred in those cells. She will now require fewer additional mutations to develop breast cancer than a woman who had a child just after puberty.

So delaying pregnancy, something associated with women's control over their own reproductive lives and also associated with higher socioeconomic status, increases the probability of developing breast cancer, as do good nutrition and a sedentary lifestyle. And as previously mentioned, modern life, including some of the best aspects of modern life, causes breast cancer.

In fact, if you think about it, virtually all of the factors that increase breast-cancer risk are specific departures from our way of life in the Paleolithic past, the time in which the major design of our bodies evolved. In those days, as we have seen, life was difficult. Women reached puberty in their late teens and got pregnant as soon as it was physiologically possible, probably in their early twenties. After giving birth, they would not have resumed cycling for several years, because they would be nursing their babies for that long. Then they would have another child and another lengthy period of nursing. Finally, the few who were still alive would undergo menopause in their mid-forties. The Australian biologist Roger Short has calculated that Paleolithic women who survived to the age of menopause probably experienced no more than one-tenth the number of menstrual cycles during their lifetime that modern women do, and the evidence suggests that they probably had a vanishingly low frequency of breast cancer.

Health-consciousness has increased the breast-cancer rate by making women more aware of the problem and more prone to self-examination and medical examinations such as mammograms. The big increase in breast cancer during the 1980s—called an epidemic by some—correlates very nicely with the purchase of X-ray machines used in mammography. And nearly the entire increase during that period was in tumors discovered early on. So even though the rate of breast-cancer occurrence has been increasing, the death rate has not. It has remained steady during all these decades of increasing occurrence.

The rise in breast cancer can be explained by changes in the length of the total reproductive period, more common delays in having a first child, increasing obesity, and increasing health-consciousness, which causes the cancer to be diagnosed at an earlier age. Ironically then, menopause in its modern context is a favorable occurrence for women's health with respect to breast cancer and several other gynecological cancers.

But that's about it for the "up" side with respect to postmenopausal health. In all other respects, menopause is anything but beneficial. Declining estrogen production just prior to menopause, and its virtual absence of production afterward, dramatically increase the risk of cardiovascular disease and death from heart attacks—much bigger health problems than gynecological cancers, because heart and artery disease kill almost 10 times as many women in Western societies as do all the gynecological cancers combined.

The fact that about the same proportion of both sexes—half—die of some type of cardiovascular disease can mislead one into imagining that men and women are really pretty similar in matters of the heart. They aren't. Men often die early from heart disease, women very rarely. By age 35, cardiovascular disease becomes and remains forever after the number-one killer of men in the United States. By contrast, it becomes the number-one cause of death among American women only after age 60.

How do we know that women's hormones protect against cardiovascular disease and men's hormones promote it? Castrate a man, and his risk of heart disease drops. Remove a woman's ovaries and her risk increases. Replace the estrogen her ovaries would have produced, and her risk drops again. Women who undergo early menopause, other things being equal, have a higher risk of heart disease. Also, the peak difference between men and women in the probability of death from heart disease occurs just prior to menopause. After menopause, the sexual difference steadily shrinks, so that by about age 90, the sexes are indeed pretty similar in matters of the heart.

Of course, if women take estrogen supplements after menopause, the whole health picture is changed.

A Brief but Necessary Digression into the Meaning of "Risk Factors"

I've been bombarding the reader, as does the popular press daily, with a litany of life's risks. Breast cancer risk is reduced by 10 to 20 percent per year the later in life that menstrual cycling begins. Late menopause and late childbearing increase the risk. Other medical reports tell us that a high-cholesterol diet increases the risk of rectal cancer by 65 percent. Decaffeinated coffee is a healthier drink than regular coffee. No, make that decaffeinated coffee is *unhealthier*. Margarine is healthier than butter. Oops, what I meant to say is that under certain circumstances maybe margarine is healthier, but on the other hand. . . .

Welcome to the hocus-pocus world of risk-factor epidemiology— the statistical study of the causes of noninfectious diseases such as cancer and heart disease. Yearly avalanches of epidemiological reports relating disease risk to everything from the believable (exposure to high pesticide levels) to the bizarre (body height, baldness) are either frightening or numbing, depending on your own personal anxiety threshold.[82] Hearing that a highly publicized finding, such as that linking saccharin to cancer, has been reversed—or that some studies find one thing, but others find the opposite—the public is likely to tune out risk factors altogether. This is not sensible, either. You just need to be an educated consumer of risk-factor research. Because risk-factor analysis also plays a large role in understanding the health or antiaging effects of various vitamins, drugs, and lifestyles, it is time to understand a bit how risk-factor analysis works, what we can believe, and what we should ignore.

What makes risk-factor epidemiology a dicey enterprise is not a lack of conceptual scientific rigor. Epidemiologists have at their command a sophisticated arsenal of rigorous statistical techniques. The problem is that humans are not lab rats. They cannot be forced to spend years shut up in identical cages, eating identical diets, living in every way identical lives except for, let's say, one group eating a high-fat diet and the other a low-fat diet. We can't *experimentally* assess the causes of disease in humans the same way we do for laboratory animals; we have to rely, by and large, on *observational* techniques. Hence, risk-factor epidemiology.

The physicist Richard Feynman summarized the attitude of scientists generally when he said that "the test of all knowledge is experiment."[83] For medicine, human experiments are called clinical trials. That is, human subjects are divided into two groups as similar as possible to each other. One group is subjected to one treatment, say, taking a placebo such as a sugar tablet every day; the other receives another treatment, perhaps an estrogen or vitamin C tablet. Ideally, neither the people giving the pills nor the subjects know who is in which group. After a time, one assesses whether the disease in question is more common in one group than in the other and by how much. Unfortunately, clinical trials are expensive and time-consuming, even if they are by far the most direct approach to gaining knowledge.

"Observational" epidemiology on the other hand, which is what

167

we more often hear about, attempts to correlate a disease with behavioral or environmental factors. It is comparatively fast and cheap and, when it correctly identifies a risk, can benefit enormous numbers of people. From such studies the effects of smoking on lung cancer and heart disease emerged, as did the relationship among high-fat diets, cholesterol, and heart disease.

One nice thing about observational epidemiology is that all you need are questionnaires, a telephone, and a computer. But here is where potential trouble begins, because correlations can easily arise by chance, or from an unintentional bias in the people included in the study, or from factors other than the one the researcher is focusing on, but which happen to be correlated with the one the researcher *is* focused on.

The two basic types of observational studies are so-called case-control and cohort studies. Case-control studies compare a group of people with a disease (the "cases") with "controls," people similar to the "cases" in as many respects as possible but without the disease. Statistical analyses then try to reveal how the two groups might systematically differ in some aspect of their lifestyle or environment.

To see the potential problems with such studies, let's consider a real example: a 1988 study that linked childhood leukemia and brain cancer to living in areas with a high density of overhead power lines. Power lines produce low-level electromagnetic fields (EMFs), and the idea behind the study was that long-term exposure to low-level EMFs somehow promoted cancer.[84] The study compared EMF exposure in homes of children who were diagnosed with any type of cancer in the Denver metropolitan area over a seven-year period with the homes of randomly selected children without cancer.

Low-frequency electromagnetic fields are produced not only by power lines but also by industrial machinery and electric household items such as razors and blankets. Therefore, determining anyone's overall long-term exposure is virtually impossible. There is also no known mechanism by which EMFs are deemed likely to cause cancer, and they have never been linked to cancer development in laboratory animals (which are much more prone to cancer than humans). Yet EMFs have become suspected carcinogens.

The Denver study estimated EMF levels in the homes of cases and controls directly with a meter and by analyzing city power-line maps. It found no statistical association with cancers when exposure was measured by meter and a modest association when calculated from power-line maps. This modest association generated headlines nationwide. But a lot of things in the study were associated with cancer, and some that should have been weren't. For instance, in addition to living near power lines, you were also more likely to get cancer if you were poor, had a less-educated father or a younger mother who smoked when pregnant, or had more automobile traffic around your house. Moreover, cancer was not associated with the number of years people had lived near the power lines or the proportion of the day spent at home near them.

It is easy to see how differing socioeconomic status might enter the picture. Controls were chosen by randomly mixing the last four digits of the cases' telephone numbers, calling these households, and asking the residents to participate in the study if they had children of the appropriate age and sex. This meant that being enlisted in the control group depended on whether there was someone at home during the day to answer the phone or, if the home had an answering machine, whether the people were willing to phone back, because they felt that this sort of study was valuable enough to participate in—all factors likely associated with socioeconomic status.

The central questions arising from the research, of course, are whether this association was real or a false alarm due to chance, and, if it was real, whether it was due to EMF levels or some other aspect of life in lower socioeconomic households. Recently, the most thorough study to date of EMFs' effects on humans appeared. A congressionally mandated review committee of the National Research Council concluded after reviewing more than 500 studies that there was no clear evidence linking EMFs to human health problems.

Uncertainty caused by a weak statistical association isn't a special problem of this particular study, which was carried out with all the rigor with which these sorts of studies can be performed. It is a problem with these *types* of studies generally. Many false associations are found. The scientific article will often include all the appropriate caveats and qualifications, but what gets reported in the press and trumpeted by the researchers' publicity department is "Coffee Linked to Pancreatic Cancer!" or "Saccharin Found to Be Carcinogenic!"— both now long-discredited associations. And as we all know, if you cry "Wolf!" often enough, soon no one will believe you.

Cohort studies consist of tracking a large group of people over long periods, trying to correlate lifestyle and environmental factors with the diseases people in the "cohort" ultimately develop. These studies are also beset with false alarms, mainly associated with the accuracy with which people remember and report various "lifestyle" factors. For instance, fat people tend to underreport, and skinny people to overreport, fat consumption. Ill people may pay more (or less) attention to and report more (or less) accurately on what they are eating or drinking; people with respiratory problems may be more (or less) honest about how much they smoke. And on, and on.

Are we stuck, then, in a hopeless morass of uncertainty, or are there guidelines that might help separate the valid from the accidental findings? Actually, there are a few sound guidelines. First, never believe a single study. Risk factors are most convincing when a variety of studies, of a variety of types in a variety of times and places, all (or almost all, because chance and poor design will always be problems) reach the same conclusion. Second, examine the magnitude of the risk. Cigarette smoking increases the risk of developing lung cancer by thirtyfold, or as it is also put, by 2,900 percent. That is a risk you can take to the bank. When you hear that something increases or decreases a risk by 20 percent or 30 percent, be skeptical unless lots of other studies have reached the same conclusion or unless there is evidence from animal studies identifying a plausible mechanism by which the effect could occur. In the absence of this sort of confirmational evidence, professional epidemiologists place little faith in risks that aren't increased at least threefold or fourfold. Finally, never believe anything in a press release or news report without verifying whether the above conditions hold.

Also, because the history of aging research is so gloriously packed with fraud, false hope, and charlatanry, these caveats apply emphatically when nostrums reported to extend health or life are considered.

Back to Health and Menopause

The previously mentioned risk factors associated with breast cancer and hormone exposure meet these conditions of rigor. All are verified by dozens to hundreds of studies. But if menopause decreases the risk of developing various gynecological cancers, and if the body does not seem well adapted to postmenopausal hormone levels, does menopause increase the risk of other diseases? Yes, it does.

The biggest problem is with heart disease. Menopause is well documented to increase the risk of cardiovascular disease. One factor that contributes to this risk is a changing blood-cholesterol profile. Women generally tend to have lower overall blood-cholesterol levels than men, lower LDL (or "bad") cholesterol and higher HDL (or "good") cholesterol. Just prior to and after menopause, as estrogen levels at first fall slowly, then plummet, women's blood-cholesterol profiles begin to change in the direction of men's. In particular, total cholesterol and LDL cholesterol increase. HDL cholesterol may or may not decrease; studies conflict on this point. But there is no conflict in opinion about how dramatic the increase in disease is during this time period.

Cardiovascular disease is not the only major health problem associated with postmenopausal hormone loss. Another major problem is osteoporosis—that is, bone thinning and loss that can ultimately lead to fractured vertebrae, "wrists" (in actuality, it is the end of the thin forearm bone, the radius, that breaks), and hips (again, it is not really the pelvis that breaks, but the end of the long thigh bone, the femur). "Hip" fractures represent perhaps our greatest unacknowledged health problem. Currently, half of all American hospital beds for trauma patients are occupied by hip-fracture sufferers. And even though people do not generally die from broken bones, the best current information is that about 20 percent of elderly women with osteoporotic hip fractures *die of complications* within one year, making osteoporosis a major killer—the 12th greatest killer in the United States—and like heart disease more deadly than all gynecological cancers combined.[85] Also, because hip fractures often disable the survivors for years, osteoporosis costs Americans billion of dollars per year.[86]

Bone, despite its stonelike appearance on your dinner plate, is a dynamic living tissue in the body—constantly being reshaped and reformed, being produced and dissolving. Both men and women lose bone as they age, because the rate at which bone is being dissolved begins to outstrip the rate at which new bone is being made. Not only does the total amount of bone decrease, but the bone that remains becomes more porous, less dense, and more fragile. Bone loss is generally far more serious for women because they typically have less bone mass to begin with and they begin losing bone somewhat earlier in life than men. But in the first five to 10 years after menopause, women experience a period of substantially accelerated bone loss. Consequently, women are much more likely than men to suffer collapsed vertebrae, broken "wrists," or broken "hips" as they age.

Again, there is no question about the role of estrogen loss in bone loss, even though the exact mechanism by which estrogen has this effect is still largely a mystery. For instance, female endurance athletes, who exercise to the point that they stop menstruating, hence manufacture less estrogen, begin losing bone mass at an age when their less active friends are still building bone. Women suffering from anorexia also stop menstruating and also lose bone at an accelerated rate. Women who have had their ovaries surgically removed begin prematurely losing bone from that time on. Indeed, the most common way to induce bone thinning in laboratory rats so that the process may be studied in detail is to remove their ovaries at a young age.

In addition to these life-threatening health problems, hormone loss at menopause also disrupts the normal sense of psychological and physical well-being in about one-third of women in Western societies to the point that they will seek medical advice for a host of new physical experiences—hot flashes, insomnia, night sweats, depression, irritability, headache, back ache, painful intercourse, loss of sex drive, problems with short-term memory, general lethargy. About another third of women report some of these experiences but at a level they consider too minor to bother seeing a doctor about. Another third of women seem to avoid these experiences altogether. We have little idea at present about what makes a woman fall into one of these categories rather than the others.

Culture plays a big role, though. Just as the incidence of many diseases varies widely across cultures, so do the signs or symptoms of menopause. Mayan women living harsh rural lives report no hot flashes or night sweats at all.[87] Only about one-third as many Japanese as American women report having ever experienced a hot flash, and Japanese women are much less likely to go to a doctor for menopausal symptoms. Japanese women also report headaches and stiff shoulders as the two most common symptoms associated with menopause, not hot flashes.[88] When I gingerly inquired about hot flashes among the older village women in New Guinea, I was treated as slightly daft (not an unusual experience during my time in New Guinea, as I now recall). Hot flashes? Was I kidding?

There are two possible explanations for this cultural variation. The first is that much of what we consider "discomfort" is cultural. A number of years ago, there was a study of what constituted pain among Western mountain climbers in the Himalayas compared with their Sherpa guides. Given the same electrical shock, with identical responses electronically recorded from their nerves, Westerners reported "pain" at a much lower level of stimulation than Sherpas. Americans and Europeans who choose to climb Himalayan peaks probably would not be considered our wimpiest representatives, either, but there you have it. So if you are a New Guinean village woman with chronic malaria, a veritable bestiary of intestinal worms, quite possibly tuberculosis, and chronic protein shortage, who has for decades performed a level of back-breaking physical labor that would kill most American men or women, it might be that hot flashes are such a trivial issue that you wouldn't notice.

But just as possibly, there are aspects of diet, exercise, and other lifestyle factors that truly alter menopausal symptoms in some cultures. Perhaps living with lower hormone levels even when still menstruating is one of these factors. Maybe eating plants that are rich in estrogenlike compounds is another.

Hormone Replacement: Does It Make Sense?

If postmenopausal life has been a common phenomenon only during the last few centuries—that is, if it is relatively new in the evolutionary scheme of things—and if women's bodies are therefore "designed" to operate best on premenopausal hormone levels throughout life, as the earlier discussion suggested, then medically replacing the hormones lost at menopause might be expected to extend life and delay or mitigate some of the specific health and welfare problems of aging women. In fact, it does exactly that. Even though women still survive life's rigors better than men after menopause, they could be doing so by a substantially greater amount.

Hormone replacement can take myriad forms. Women can take estrogen alone, combine estrogen and progesterone in different ways and different doses, take pills, or use pellets placed under the skin, nasal sprays, or skin patches. Also, there are a variety of synthetic hormones to choose from and dozens of new types of hormone treatments under development.

Most studies of hormone replacement have focused on estrogen, because administering estrogen alone is the oldest type of replacement therapy. Observational studies on survival in women taking estrogen replacement have varied widely in a number of details— the ages of women in the study, the length of time they used estrogen, what was known about the dose and duration of use, how many years the women were tracked—yet in spite of these differences, virtually all of the studies to date found a reduced risk of death among estrogen users. By far the most comprehensive of these studies[89] was of long-term (average of 17 years) estrogen users who had begun use soon after menopause. This study followed women for more than 25 years, and because these women were members of a large health maintenance organization, researchers had access to all their medical records and, from pharmacy records, even knew the precise estrogen doses the women had taken over that time. The yearly death rate among estrogen users turned out to be about half that of nonusers.

This shouldn't be surprising because it is so well established that hormone replacement (either estrogen alone or combined with a progesteronelike hormone) pushes cholesterol profiles back in the premenopausal direction and substantially lowers the overall risk of death from the largest killer of elderly women, cardiovascular disease. Interestingly, though, evidence from a recent *clinical trial* of hormone replacement as part of the U.S. National Institutes of Health's Women's Health Initiative suggests that these changes in cholesterol profiles account for no more than half of estrogen's protective effect on heart disease.[90] The other half may have to do with its effects on body fat, blood pressure, elasticity of arterial walls, or a host of other factors. We simply don't know at present. However, we do know that a having a good cholesterol count does not confer immunity from heart attack or mean that hormone replacement will not help protect you from one. Thus, because cardiovascular disease kills about 10 times as many women as all gynecological cancers combined, hormone replacement makes sense from a purely statistical point of view, unless it increases gynecological cancer rates massively. And as we shall see below, there is little suggestion that this is true.

Just as overwhelming as its effect on cardiovascular disease is hormone replacement's effect on bone loss: Users suffer many fewer fractures from osteoporosis than do nonusers if the hormone therapy is begun relatively soon—within a few years—after the ovaries stop functioning and is continued for the long term. In fact, a recent report from the same clinical trial that reported such a substantial reduction in deaths from heart disease with hormone replacement demonstrates that hormone replacement not only slows bone loss, it actually reverses it.[91] From what we know now, if begun later in life, say in a person's seventies, or used only sporadically, hormone therapy seems to have little effect on bone loss and fractures.[92] The importance of continued use must be emphasized, because only about one-third of women who are prescribed estrogen actually take it as pre-scribed. The rest never have the prescription filled, stop taking the estrogen within the first few years, or take it only sporadically, per-haps when symptoms such as hot flashes become particularly notice-able.[93] The major benefits of hormone use apply only if the hormone is taken regularly and for long periods.

One big difference between the risk of osteoporotic fractures and death from cardiovascular disease is that it is much simpler to deter-mine your risk of suffering a fracture. That risk is a straightforward function of bone mass and density—easily (although not cheaply) and painlessly measured using a variety of modern techniques that go by tongue-twisting names such as single-photon absorptiometry, dual-photon absorptiometry, dual-energy X-ray absorptiometry, and computed tomography (the so-called CAT scan). Standard X-rays can detect fractures that have already occurred, but are not sensitive enough tools for assessing one's future risk of fracture until that risk becomes very high. So a thorough examination of your skeletal sys-tem can let you know with fair certainty whether you are at risk for osteoporotic problems. The risk of serious cardiovascular disease, on the other hand, is very difficult to assess because it really encom-passes many different problems in the structure and function of the heart itself or of the blood vessels. So it isn't surprising that we have probably all heard of friends, neighbors, or celebrities who have dropped in their tracks from a heart attack within days of being given a clean bill of health after a thorough physical examination. Remem-ber Reggie Lewis, the Boston Celtics basketball player, who fell dead in a playground game within weeks of undergoing a more thorough heart examination than most of us will ever have?

In addition to reducing the risk of these life-threatening prob-lems, hormone-replacement therapy indisputably alleviates such non-life-threatening discomforts of menopause as hot flashes, insomnia, and depression in more than 90 percent of users.[94] There may be additional benefits of hormone use, although this information is still preliminary and far from fully established. For instance, a few studies have indicated that estrogen replacement is associated with reduced rates of rectal and colon cancer.[95] Estrogen therapy may even protect aging women against the development of Alzheimer's disease,[96] and reports just now beginning to appear suggest that patients who already have Alzheimer's may experience improvements in their symptoms through estrogen therapy.

With all these apparent benefits, why isn't estrogen-replacement therapy automatically recommended after menopause in the same way that doctors recommend blood-pressure-lowering drugs to hypertensives? After all, if it came from plants and had an exotic name, any drug with this sort of clinical record would be heralded as a miracle drug for slowing the aging process. Why is it still controversial in the public at large, and why are women so hesitant to take estrogen even after it's prescribed for them? One answer is that some women experience unpleasant side effects, such as nausea and headaches, when they take estrogen. And depending on the type of hormones one is taking, some women will experience vaginal bleeding, a sort of pseudomenstruation.[97]

But the big reason that more postmenopausal women don't take hormone replacement is a fear of gynecological cancer. If the amount of natural lifetime exposure to estrogen and progesterone is well established to increase one's risk of developing various gynecological cancers, wouldn't the expectation be that artificially increasing this exposure by taking estrogen postmenopausally would similarly increase one's risk? The short answer is yes, that would be the expectation, but expectations are not necessarily fulfilled. Hormones absorbed through the digestive tract or skin will not necessarily have the same effects on reproductive tissues as hormones that are produced locally by the ovaries. Also, the timing, duration, amount, and combinations of hormones taken can be manipulated in ways that our bodies never do.

We can see how appropriate manipulation of hormones might reduce their potential cancer-causing effects if we consider how hormones influence cell division (and therefore the risk of mutations) in different tissues. Cancer-prone breast cells divide most rapidly during that phase of the menstrual cycle when *both* estrogen and progesterone occur at high levels, whereas the cancer-prone cells of the lining of the uterus (the endometrium) divide most rapidly during that phase when *only* estrogen is elevated. High progesterone levels seem to inhibit cell division in the endometrium even when estrogen levels are high. This is important for understanding the effect of hormone replacement on gynecological cancers, because during the 1970s, when estrogen was prescribed in considerably stronger doses than today and when hormone pills almost always contained only estrogen, there were reports of endometrial cancer rates increasing in hormone-replacement users as much as tenfold. As a result, it became commonly accepted that hormone replacement medications should include *both* estrogen and progesterone to reduce the risk of endometrial cancer. Indeed, women taking these "combined" hormone supplements today generally exhibit lower rates of endometrial cancer than nonusers.

What about breast cancer? If estrogen and progesterone together stimulate high levels of cell division in breast tissue, shouldn't combined hormone treatment be expected to increase breast-cancer risk? Again, maybe, maybe not. Hormone supplements can utilize synthetic hormones that do not last as long in the body as natural hormones. Also, doses can be manipulated to keep them as low as possible to have the desired effect. Prescribed hormone dosages have been dropping over the past 20 years. In any event, the observational studies on hormone replacement, whether estrogen only or combined, and breast-cancer risk are decidedly mixed. Most individual studies fail to find any increased risk with hormone replacement. Of five recent studies that combined the information from many individual studies, two concluded there was no increased risk; the other three found a small risk. The most recent large case-control study found no increased risk for either estrogen-only or estrogen and progesterone combined.[98]

It is important to emphasize several aspects of the state of knowledge at present. First, even if one assumed that only the highest cancer risks reported in recent studies were valid, the overall reduction in deaths from cardiovascular disease would far outnumber the increase in deaths from gynecological cancers if every postmenopausal woman were on hormone replacement. Second, we will never learn the optimum hormone doses and combinations if we are forced to rely only on "risk-factor" epidemiology for our information. Large long-term clinical trials such as the Women's Health Initiative need to be continued and expanded if we wish ever to fully understand these issues. Finally, there is a major lack of experimental animal research to address these issues, because rats and mice—the experimental animals of choice in medical research—do not undergo anything similar to human menopause. There is a tremendous resistance in the medical-research community to use any animals but rats and mice, because we now know so much about these laboratory rodents that we can dissect their inner workings with remarkable sophistication. However, new animals are needed to study some issues, including menopause and aging generally. Perhaps we even need to examine the physiology of such animals as pilot whales and killer whales, with their long evolutionary history of menopause—a menopause in which evolution should have molded their bodies to thrive on reduced hormone levels.

Or do we really need more animal research and more experiments? Maybe the answers to slowing aging and preventing disease are already here. You can hardly walk through a bookstore or look through a magazine these days without coming across the "antiaging" this or the "life-extending" that. There have been books published on how to live to be 120 or 180 years old, take your pick. There currently exists an American Academy of Anti-Aging Medicine, which promotes itself as scientifically respectable, and which holds yearly meetings attended by thousands, including lots of press. But antiaging treatments have been available for centuries. All you had to do was lie between two virgins, or sit in baths of the proper temperature, or eat algae from the right lake. In the 1920s (and probably even today), famous people were paying big bucks for antiaging injections of macerated dog, goat, or monkey testicles. I said earlier that the best way to increase your longevity was to live in a place that doesn't keep reliable birth records. Is that really still true, or with our modern knowledge about oxidative damage, free radicals, and browning damage, and our understanding of how cancer and heart disease develop, do we already have antiaging treatments available to us? The next chapter investigates exactly that question.

10

Slowing Aging and Extending Life

Remedies and Expectations

*Danger lies not in what we don't know, but in what
we think we know that just ain't so.*

ANONYMOUS

I'm going to pluck randomly from my long shelf of books explaining how you go about aging more slowly. There. This one happens to be called *Anti-Aging: Secrets and Strategies*, written by Anonymous in slightly breathless but perky prose and published in 1993 by M Press (of no apparent address). The first thing I notice is that the book's title page contains a lengthy disclaimer written in densely technical legal jargon. But its meaning is straightforward—it protects the author and publisher from any liability associated with "damage or loss" caused by advice the book offers. Yet the advice seems no more controversial—or revolutionary—than the sensible platitudes perennially invoked by mothers and scout leaders. So among the book's "Top Ten Anti-Aging Secrets," you find the following: don't smoke, eat right, exercise, drink alcohol moderately, get enough sleep, have regular medical checkups. Yes, there are some howlers, too. Among "Secrets for a Youthful Appearance" is the suggestion to consider cosmetic surgery. Aroma therapy and recreational shopping are also recommended. None of this advice is bad. I have nothing against pleasant odors, and although I'm not a world-class shopper, I think I understand its psychological appeal. What the publisher's lawyers were worried about, I suppose, was that none of this advice has anything at all to do with delaying aging. The perfect Boy Scout will still be middle-aged in his forties and fifties and old in his seventies and eighties. Remember, aging is not just developing diseases, it is generalized deterioration. Postmenopausal hormone therapy and certain diets (low fat, lots of fruits and vegetables) are known to reduce your chances of developing specific diseases, such as heart disease and cancer. A large group of people taking hormones or eating such diets would on average live a bit longer than they would if they were not watching their diet, because so many of us die of heart disease and cancer. But we couldn't really say that these treatments had retarded aging per se, unless they also slowed the rate at which their practitioners lost their hearing and sprint speed and knack for picking up new languages. Also, treatments that truly retarded aging would not only increase average longevity, they would also increase maximum longevity—that is, how old their longest-lived practitioners lived to be. Nothing to date has been demonstrated to live up to this expectation.

In the following pages, as I discuss the promise and prospect for slowing human aging, I will purposely not address the very real philosophical issue of whether retarding the aging process is socially, politically, or ethically desirable. Those questions are not the province of science, and scientists, in spite of their sometimes unfortunate bent toward hubris, have no unique insights to offer on that score. It does seem compellingly apparent, though, that regardless of the social desirability of slowing aging, if science uncovers therapies that can do it, those therapies will be employed. This is one genie that has no chance of being put back in the bottle.

Let me emphasize one thing clearly at the outset so that whatever I may say later will not be misunderstood. *There are currently no diets, no vitamin or mineral or hormone supplements, no attitudes, and no behaviors or lifestyle choices that have been demonstrated to slow aging in humans!* Some currently advertised antiaging treatments and behaviors consist of sound health tips, even though they may have nothing at all to do with slowing aging. Others are unproven even with respect to health, and still others can be dangerous. But given what we know about life's damaging processes, there are some likely candidates for real antiaging treatments for humans in the not-too-distant future. In the rest of this chapter, I will sort out the various claims, assess the existing evidence, and discuss reasonable promises offered for really slowing aging in humans.

Wishful Thinking

Let's dismiss the truly ridiculous ideas first. Among the multitude of putative antiaging products on the market, so-called natural products or anything from Asia seem to have special cachet. One of the more interesting of these antiaging products is royal bee jelly, which does in fact prolong life enormously—in honeybees. Honeybee workers live for a few weeks, whereas queens can live for many years. There is no genetic difference between workers and queens. Any female in an undeveloped larval state has the potential to develop into either. Whether a larva actually does develop into a worker or queen depends largely on what it is fed. A steady diet of royal jelly produces long-lived queens. Larvae fed more typical bee rations develop into short-lived workers.

The rationale for hoping that what extends life in bees that typically live a few weeks will also extend life for humans is rather obscure. Putting bees in a refrigerator will also extend their lives enormously, yet we don't find refrigeration therapy for long life offered in the magazines. Perhaps the reason is that refrigeration therapy could be so quickly and decisively disproved, although finding sufficient volunteers to run the critical experiment might be a problem. Quite analogously, whether termites develop into workers or queens depends largely on certain aromas, but I haven't yet heard of termite aroma therapy for long life. Maybe I'm just ahead of the curve on this one. Interested venture capitalists should feel free to contact me, and we will work out the legal and financial details and a marketing strategy.

All kidding aside, there is a long and inglorious history of similarly bogus antiaging therapy that deserves to be laughed at long and hard and bitterly (for those of us who didn't cash in). If it wasn't injection with emulsified goat testicles, it was eating special algae or meditating in the proper position or receiving injections of dried cells from fetal pigs, sheep, or rabbits, or drinking Chinese herbs or Novocain. This last one, Novocain, the same chemical (procaine hydrochloride) used as a dental anesthetic, is still available in Nevada. Developed by a youthful-looking Romanian physician, Ana Aslan, 40 years ago and marketed in a form that is chemically stablized and called Gerovital H3, this antiaging therapy has less than zero scientific support. It has negative support. That is, all experiments done on it outside Romania agree that it has no effect on aging whatsoever. Its only known effects are as a mild antidepressant that causes allergic skin reactions in some people. This didn't prevent Dr. Aslan from becoming wealthy enough by treating the gullible rich to open her own research institute and become a national heroine. It also didn't prevent her from dying in 1988 at the age of 91—not Vilcabamban longevity, but not bad by east European standards.[99]

The rest of the therapies I will discuss at least have a scientific rationale and deserve further investigation.

Cutting Calories, Living Longer: What We Know and What We Would Like to Know

There is no controversy at all about the fact that restricting the amount of food eaten by *laboratory rodents* slows their aging rate. In fact, food restriction is *the only* proven method of reliably slowing aging in laboratory rodents. No one—repeat, no one—knows why food restriction has this effect or how it works. Also, no one knows whether it will have a similar effect in humans. Such a lack of knowledge doesn't stop researchers studying food restriction in animals from having strong opinions on the topic, though.

Before assessing the cogency of these opinions, we need to be specific about how rodent food-restriction experiments are done in the laboratory. Usually, the amount of food eaten by laboratory rats and mice is not measured. They typically live their entire lives in shoebox-size cages and are fed *ad libitum* (usually abbreviated as *ad lib*), meaning that they always have excess food available and so can eat as much as they wish. If you compare rats or mice fed this way with identical animals that have eaten much less—usually 60 to 70 percent, but in mice occasionally as little as 35 percent of the calories eaten by animals fed *ad lib*—the dieting rodents will develop fewer tumors and other diseases than their well-fed kin. They will maintain their youthful energy, immune system, and memory longer. They will suffer less oxidative damage to their tissues. Their tendons and ligaments will stiffen more gradually. They will survive and recover from surgery better. They will be more resistant to chemical carcinogens. Their lives will, on average, be extended 25 to 40 percent or more, and the longest-lived restricted individuals will be at least that much longer-lived than the longest-lived fully fed animals.

This very real antiaging effect of food restriction doesn't seem to depend on whether the calories are cut from major nutrients such as fats, protein, or carbohydrates. The important thing seems to be only that sufficient vitamins and minerals are supplied as total calories are reduced. By and large, the antiaging effect is larger the greater the magnitude of caloric restriction (short of starvation) and the greater proportion of an animal's life it eats reduced rations.

The reason that researchers have had so little luck in determining what causes this antiaging effect is that so many things change physiologically when food is restricted. Some changes occur directly in those general damaging processes now suspected to be fundamentally involved in aging. So, for instance, food-restricted animals produce fewer free radicals and have less glucose dissolved in their blood than their fully fed counterparts. It isn't surprising therefore to find that both oxidant- and glucose-based damage is mitigated by food restriction.[100]

But a number of other changes occur, as well, and those changes may differ somewhat between rats and mice. For instance, both species exhibit delayed puberty if food is restricted as soon as they are weaned from their mothers, but adult rats show only a reduced reproductive rate, whereas reproduction ceases altogether in mice. These reproductive effects associated with delayed aging are completely consistent with what evolutionary aging theory would predict. Also, body temperature falls in mice, but not significantly in rats. These differences between the two species, both of which exhibit delayed aging in response to food restriction, help sort out which physiological changes may be crucial to that response. Therefore, the fact that food restriction fails to lower body temperature in rats as it does in mice indicates that reduced body temperature is not *required* for the antiaging response.

But there are many more similarities than differences between the species. Almost all hormone levels are reduced in both species, as are rates of cell division. The one thing that is not reduced in either species is metabolic rate per cell. More precisely, there is decreased metabolism when restriction is first begun, but over a period of a few weeks to a few months, depending on the age of the animals, physiological adjustments are made that finally return metabolism to at least the same level—potentially a bit higher. Raymond Pearl may be rolling in his grave, but food-restricted animals are not living at a reduced rate. They have somehow reduced the production of oxidants without burning less oxygen.

To some researchers, a substantial surprise has been that certain stress hormones actually *increase* in food-restricted rodents. To anyone who has ever gone on a serious diet, this is probably less of a surprise. Dieting is apparently no fun for mice, either. But the conventional medical wisdom has always been that stress is uniformly unhealthy. Excess stress has in fact been linked to disrupted digestion, stiffened arteries, stunted growth, devastated fertility, and crippled immunity. There may be no more common (and, as it turns out, probably erroneous) bit of cocktail-party medical lore than that the one sure road to a premature heart attack is to be a stressed-out, type A, overachieving personality.[101] How then do food-restricted animals not only escape the damaging effects of increased stress, but actually appear to thrive on it? At this point, no one knows. It may be that food restriction stimulates such a wide-ranging array of protective mechanisms that aging is slowed *in spite of increased stress*, or some researchers now believe that some moderate level of stress may be beneficial.[102]

The idea that a little stress may be beneficial is not new. It is called *hormesis* and grew out of observations in the 1940s and 1950s that low doses of otherwise toxic substances such as nuclear radiation and insecticides frequently increased the growth rate in animals. We can thank these observations for the rash of dreadful horror movies in the 1950s in which accidently nuked ants or spiders or lizards grew to the size of buildings and required the military in order to be subdued. Some additional experiments in the 1960s seemed to suggest that the moderate stress of irradiation or electric shock or cold exposure would increase longevity, too.[103] After having been generally dismissed for the past couple of decades, this idea is now getting a fresh look.

However it works, the food-restriction phenomenon must always be kept in mind when evaluating other putative remedies for aging. Any remedy that reduces food consumption in rats or mice by, say, making food less palatable or suppressing the appetite in any way will inevitably delay aging if it isn't too toxic in other respects. Surprisingly, in spite of this obvious caveat, most experiments testing other antiaging therapies have failed to measure food consumption.

Naturally, the $64,000 question is whether the antiaging effect of food restriction operates in people the same way it does in laboratory rodents. One influential U.S. senator in the mid-1980s thought the question worth considerably more than $64,000 and personally saw to it that $1 million suddenly became available to begin a food-restriction experiment in monkeys. Because humans are much more closely related to monkeys than to rodents, and because monkeys live a more humanlike period of decades rather than a couple of years, it was reasonably thought that a crucial test of food restriction's antiaging properties for humans would be its effect on monkeys. This logic is sound. However, precisely because monkeys live for decades even when they eat like pigs, definitive results from these experiments will not be available any time soon.

What we know from three ongoing monkey studies of food restriction is a mixed bag of similarities to, and differences from, the laboratory rodent research. First, there seem to be differences in the way different monkey species respond. After three years of food restriction, rhesus monkeys, like laboratory rodents, were lighter and leaner than control animals. Squirrel monkeys, on the other hand, were not.

Focusing only on rhesus monkeys, which seem to respond most like rodents, food restriction delays puberty and lowers blood glucose and insulin levels (as happens in laboratory rodents). On the other hand, food-restricted rats tend to show increased movement inside their cages, whereas rhesus monkeys show decreased movement. Monkeys' body temperature also falls by about 1° Fahrenheit in monkeys—less than the decline in mice, but more than in rats. Whether the metabolic rate is altered by long-term restriction isn't yet clear. There simply isn't enough information on monkeys yet to tell us much. But there will be. Check back in 30 years or so, assuming the studies continue to get funding.

In the meantime, is there anything sensible to be said about the chances that food restriction will delay aging in humans? The first thing to think about is exactly how rats and mice compare with humans, as well as with each other. As mentioned earlier, virtually all medical research is performed on laboratory rats and mice not because these species are particularly close relatives of people or are exceptionally similar to humans physiologically, but because they are cheap, easy to take care of, quick to reproduce, and short-lived. They are, however, close relatives of each other. So observing that both rats and mice exhibit the antiaging effect of food restriction is not a strong argument that it will be a general phenomenon.

Humans, mice, and rats do share the fact of being mammals and therefore will share many similar biological traits, which is why laboratory rodents can be used for preliminary screening of how therapeutic, toxic, or carcinogenic new drugs and environmental contaminants might be. However, any medical-research specialist can give you chapter and verse on how rats and mice differ from humans in her particular specialty, be it heart, kidney, brain, or muscle function. Rodents have vastly different dietary requirements than humans.

They are poisoned by some chemicals that are harmless to humans, and vice versa. Rats and mice do not develop the neuritic plaques characteristic of Alzheimer's disease as they age, while humans, monkeys, and dogs do. Lab rodents have vastly different reproductive systems from humans, and minuscule hormone levels in comparison. Unlike humans, rodents manufacture their own vitamin C rather than getting it from their diet. They prefer to sleep during the day and boogie at night. Rats even respond differently to short-term starvation than humans. If you starve a laboratory rat for a day, it can run farther and longer than if it were fully fed. Starve humans for the same amount of time, and exercise endurance plummets.[104] But certainly the fact that a drug or other treatment has a certain effect in rats or mice *suggests* it might very plausibly have the same effect in humans.

Another significant factor to bear in mind, though, is that laboratory rodents themselves are nothing like the rodents found in nature. Laboratory animals have been domesticated. That is, they have been selectively bred for hundreds of generations for traits amenable to laboratory life, such as not biting their handlers, not trying to escape each time the cage lid is lifted, and not going quietly mad during an inactive life confined in a shoebox-size enclosure with the same three other individuals.

They have also been bred to eat a lot so that they grow as quickly, and reproduce as quickly and copiously, as possible, thereby making as much money as possible for commercial rodent suppliers. This type of selective breeding is now known to have shortened laboratory rodents' lives, probably by making them larger. For several decades, as better and better food and housing were developed, longevity of laboratory rodents increased dramatically, but as these advances in husbandry slowed in the 1970s, longevity began declining. Taking the standard laboratory rats (Sprague-Dawley strain) as an example, 60 to 70 percent of them lived two years or longer when fully fed in the 1970s, but by the 1990s, only 10 to 20 percent lived as long as two years. During that same 20-odd year period, the average rat has also increased by about 30 percent in body weight.[105] An unavoidable implication of this discovery is that laboratory breeding practices seem to have produced rodents prone to eating considerably more than is healthy for them. Put today's bigger rat on a diet and it lives just as long as its 1970s great-great-great-great-great-grandad. Could this mean that we are really observing the effects of overeating by *ad lib* fed animals rather than undereating by food-restricted animals?

Wild rodents, if reared in the laboratory, given excess food, and confined to a shoebox cage, will eat less, exercise more, grow more slowly, reach puberty later, and reproduce less rapidly than a domesticated rodent. In fact, wild mice fed *ad lib weigh less than food-restricted* laboratory mice. Wild mice are not long-lived in the laboratory, however. This has generally been attributed to the massive stress they experience living in confined and crowded laboratory conditions. These various factors have led at least one noted gerontological researcher, Leonard Hayflick, to believe that the food-restriction phenomenon is due to laboratory rodents having been bred for unhealthy gluttony, and that the so-called restricted diet more closely mimicks the way rodents eat in nature—the diet on which their physiology has evolved to thrive.

In strict terms, this is probably not the case, because the number of calories that seems to slow aging most dramatically inhibits reproduction—shuts it down entirely, in fact, in mice. If mice in nature ate so little, no one would need to beat a path to the door of a man who developed a better mousetrap. Mice would long since have become extinct. But food-restricted mice may be eating something closer to a wild mouse's natural diet. Just enough less, perhaps, to inhibit reproduction, thus retarding aging by redirecting the body's resources to survival instead of reproduction.

None of these differences between wild and domesticated rodents necessarily suggests anything one way or another about the probability that food restriction will work in humans, but it does make one wonder what a restricted human diet would look like. This issue is, in fact, critical to translating the rodent experiments into human terms. Are humans more analogous to laboratory rodents or wild rodents, and what is the human equivalent of an *ad lib* rodent diet? It isn't exactly hot news that overeating compromises human health and shortens life. Are the rodent experiments exercises in the control of overeating, or are they really about something else? Is food restriction to the point of slowing down aging a separate phenomenon?

Even if food restriction is something different from controlling gluttony, how would we go about identifying the right level of restriction for humans? After all, designing a standard diet for a laboratory rat or mouse is easy. All individuals are about the same size, and being confined to a small cage for life without much opportunity for exercise, they don't differ enormously in the amount of energy they expend. But designing a standard human diet is much more difficult. Humans differ dramatically in body size, metabolism, and activity levels. Elite endurance athletes stay slim on diets containing more calories than most of us could choke down. Poor laborers from India can work hard on what for Americans would be a starvation diet. Chinese people in China eat more than 500 calories per day more than Chinese immigrants in North America, yet weigh less and are leaner.[106] So what caloric standard might we use to restrict from?

Perhaps the appropriate approach would be to focus on leanness rather than calories eaten. Nutritionists generally use something called the Body Mass Index, abbreviated BMI, to judge relative leanness or obesity. Basically, the BMI is a ratio of weight to height (technically, weight in kilograms divided by height in meters squared). The important thing is that the larger your BMI, the heavier you are for your height.[107] The latest U.S. government weight guidelines recommend BMIs for adult women that range from 21 to 27.[108] For an average-size five-foot-four-inch woman, this means weighing between 122 and 157 pounds. In many Third World countries, by contrast, BMIs *average* 18 to 19 and are as low as 15 to 16 (87 to 93 pounds for the same five-foot-four-inch woman) in many healthy, hardworking people. Should the human restriction target be 30 percent below the current government standards for pudgy Americans? Or should it be calibrated to Third World values, which probably more closely represent the BMI we evolved with. Or should it be somewhere in between?

This isn't a trivial issue if antiaging dietary recommendations are ultimately to be made. We know you can eat too little to remain healthy, as people with advanced anorexia nervosa (BMIs of usually 15 or below) have taught us. We can't just assume that less is better. It isn't even clear that less is better in humans within "normal limits" in a country such as the United States. One bit of evidence arguing against food restriction as a way to delay aging in humans is that there is no medical consensus that being exceptionally lean is even as healthy as being a bit more plump.[109] A number of studies, including those by insurance companies with a vested interest in the health of their clients, have reported higher death rates among both people who were somewhat thinner than average as well as among obese people.

Even the latest, most complete information doesn't clearly suggest that leaner is better. In 1995, the *New England Journal of Medicine* published a report that finally seemed to indicate that slender people were on the whole healthier. It was the most comprehensive epidemiological study of health and relative leanness to date. A cohort study, it included more than 100,000 American nurses and corrected, as previous studies hadn't, for that meretricious leanness associated with smoking and other health problems. Its widely publicized conclusion was that indeed the leaner you are, the lower your odds of dying at any specific age. However, an examination of the study itself, as opposed to a reading of its conclusions as published in the popular press, reveals that the difference in death rates for five-foot-four-inch women weighing between 110 pounds and 155 pounds (BMIs of 19 to 26.9) is so small as to be statistically undetectable. Death rates rise significantly only above the point of true obesity.[110]

I'm not sure whether it is comforting or not to know that the problem of what is the proper baseline diet from which to restrict is confusing even to the professional researchers performing the food-restriction experiments on monkeys. The problem is that captive monkeys, like captive lions or dogs or almost any captive animal, tend to become obese when fed all they want and are not able, or motivated, to exercise. Not surprisingly, each of the monkey studies has defined the amount of food eaten by their nondieters, or control animals, a bit differently. The food-restricted animals in one study, in fact, weigh more than the nonrestricted group in another, emphasizing how difficult it is to translate rat and mouse studies into monkey, much less human, terms.[111]

So belief about whether food restriction will delay aging in humans seems to depend on one's hunches, buttressed by appropriate anecdotes. Those who don't believe in food restriction as an antiaging therapy for humans like to point to the epidemiological studies of increased or no difference in mortality rates associated with low BMIs and to note that there are many people in poor countries throughout the world who live on even less than 60 percent of a normal American diet. Some of these people, like my friends in remote Papua New Guinea, even have a putatively healthy diet that is rich in fruits and vegetables and low in animal fat. Of course, they are anything but long-lived. Quite the reverse. Admittedly, they have many chronic health problems that we don't have—tuberculosis, malaria, intestinal parasites galore—but you might expect that somewhere, at some time, at least a few people eating such diets would have escaped enough of these problems to survive the 150 or so years that the rodent experiments suggest they should.

On the other hand, believers in food restriction malign the American diet and claim that mortality doesn't decrease with low BMIs because of the quality of food eaten, not its quantity. They also like to cite the people of the Japanese island of Okinawa. Of all the places with reliable records, Okinawa has the highest fraction of centenarians in the world. Okinawa has more than quadruple the proportion of centenarians as Japan as a whole (18.5 versus 4.5 per 100,000 population) and about half again as high a proportion as the next longest-lived Japanese prefecture.[112] Okinawans also eat less than most Japanese. Children eat only about 60 percent of the calories recommended for the country at large, and adults consume 20 percent less than the national average. Okinawans are also smaller and lighter than the average Japanese. Superficially, one could draw a lot of parallels with the food-restricted rodent studies. But of course Okinawans are likely to differ from other Japanese in a large number of other aspects of their lives than just calories eaten. Okinawans themselves credit their climate and propensity for hard work. They may even have an unusual genetic constitution compared with the rest of Japanese.[113] Admittedly, these anecdotes are not very compelling one way or another, but, on the other hand, good information about this subject is hard to come by. The best thing we have to go on besides anecdotes, and it isn't much, may be the inadvertent food restriction experiment done in Biosphere 2.

Very likely the most ardent, and certainly the most visible, of the True Believers who are convinced that existing evidence is already so compelling that there is little question that food restriction will slow aging in humans is Roy Walford. Walford was the medical doctor in Biosphere 2, a three-acre air-conditioned greenhouse constructed in the Arizona desert.

Walford is one of the truly unusual characters in aging research, or any scientific research for that matter. Looking a bit like a gaunt Yul Brenner, emitting a quietly compelling confidence that he is right, Walford is a prolific immunologist who has also found time during his life to participate in avant-garde theater, write (with coauthor Richard Weindruch) the most complete review of the effects of caloric restriction on rodents, produce a string of popular books on diet and aging, and most recently spend two years immured with seven other people inside Biosphere 2.

Demonstrating that his link with living theater isn't yet dead, Walford and the other Biosphereans, garbed in *Star Trek*-like uniforms, entered the airlock sealing them away from the outside world in September of 1991 to the cheers of Hollywood celebrities, costumed dancers on stilts, and Indian chanters and fire jugglers. The Biosphere itself contained a small replica of a rain forest; patches of desert, savanna, and ocean; and land set aside to farm. Although its scientific aims have never been completely clear to me, the Biosphere's premier objective was to be totally self-sustaining. There was to be no exchange of air or food or other materials between it and the outside world for the entire two years of the project.

The Biosphere project was, and remains, controversial. Whatever its participants were trying to demonstrate, it is clear that things quickly began to go wrong. Someone was injured and had to leave the Biosphere for treatment. Oxygen and carbon-dioxide levels got dangerously out of whack and had to be adjusted to prevent the Biosphereans from suffocating. Most notably, in retrospect, it turned out that the Biosphereans weren't the farmers they thought they would be and couldn't produce nearly as much food as they had projected. In a classic example of being handed lemons and proceeding to make lemonade, Walford discovered himself in the perfect situation to begin a caloric-restriction experiment on humans.

You can't help but respect Walford. He believes in the honorable, but alas vanishing, medical tradition of performing experiments first on yourself. So he had already been restricting his own diet for years, eating all he wanted every other day and fasting in between—the same feeding schedule as his calorically restricted mice. How much he was restricting himself by this technique isn't clear, but the word in the gerontology community was that on the days he was eating you risked losing a hand if you reached for the food at the same time he did.

Having realized that they were not going to be able to grow the 2,500 calories per day that they had expected, the eight Biosphereans, ranging in age from 28 years to the 67-year-old Walford and in weight from 116 to 210 pounds, decided that they would divide the available food in a ruthlessly egalitarian fashion. Each would receive the same food ration regardless of his or her weight or degree of hunger. It turned out that this averaged out to be a bit less than 1,800 calories per day, more than a poor Indian laborer takes in, but far below the American norm. During the next six months, as they labored long and relentless hours farming, the Biosphereans each lost between 11 and 46 pounds. The average BMI fell from 23 to just under 20. Even the previously restricting Walford lost 20 pounds.[114]

Their diet was a nutritionist's dream. It was largely vegetarian. They ate six kinds of fruit (bananas, figs, guavas, lemons, papayas, and kumquats), five grains (oats, rice, sorghum, wheat, and corn), split peas, peanuts, several types of beans, about 20 types of vegetables and greens, and potatoes. There was also a smattering of animal food—milk, eggs, and yogurt, and a bit of meat from pigs, goats, chicken, and fish. They also took daily vitamin and mineral supplements.

Whether this diet slowed their aging rate is impossible to say. To my eyes, Walford did not look particularly rejuvenated upon reemergence from the Biosphere. He was the oldest of the Bi013phereans by more than 25 years to begin with and seemed drawn and rather listless compared with his normal energetic self, although he said he felt fine. Whatever else it may have done, the Biosphere diet did produce changes in the Biosphereans that might be construed as consistent with good health. As a group, their blood glucose fell by 18 percent (which one hopes reduced the rate of browning damage). Their blood-pressure and cholesterol levels also dropped. Blood triglycerides, too, fell in the men, but increased among the women. Signs consistent with good health may not necessarily mean good health, though. When my malaria returns periodically, my blood pressure and cholesterol drop nicely, even though I feel at least half-dead. Similarly, all the changes seen in the Biosphereans are also consistent with incipient protein malnutrition.

Quite surprisingly, this low-calorie diet did not affect menstrual cycling in the women, as often happens during rapid weight loss and inevitably in food-restricted rodents. Perhaps even in their emaciated state, the female Biosphereans were not dieting to the extent that we force on laboratory rodents. For fun, I looked into how lean rats became in one fairly typical food-restriction study. In the restricted animals, the BMI fell by 22 percent, compared with a reduction of about 14 percent in the Biosphereans.

It's also impossible to determine the extent to which these various changes had to do with food restriction rather than other aspects of their confined lifestyle. At times, oxygen in the Biosphere fell to that equivalent to living at an altitude of more than 17,000 feet. Also, carbon-dioxide levels skyrocketed, as did other atmospheric gases to the point that could have interfered with vitamin synthesis by the Biosphereans. And vegetarian diets by themselves typically lower blood pressure and cholesterol even when there is no attempt to restrict calories. Regular exercise also decreases blood pressure and cholesterol, and the Biosphereans were getting plenty of exercise.

Although information is not yet in hand to make any informed conclusion about whether food restriction delays aging in humans, it isn't too soon to conclude that even if it were proved to work tomorrow, humans would not be living longer any time soon because of that knowledge. We as a species seem to have very little control over our eating habits or the weight-loss industry would not be such a spectacular perennial moneymaker. For all the well-known and widely advertised adverse effects of obesity on health, humans in countries where high-fat food is freely available continue to fatten like feedlot beef. People have difficulty restricting their own calories enough to ward off obesity, much less restricting themselves to 60 to 70 percent of the normal caloric intake. About one-third of Americans are overweight—that is, 20 percent or more above "desirable" weight. Average weight has increased by eight pounds just during the past 15 years as we have been bombarded with reports of how unhealthy it is to be obese.[115]

There are also periods of life when reducing calories is clearly not a good idea anyway. Excessive leanness in adolescence stunts growth, delays puberty, and, in girls at least, can lead to premature bone thinning. So even if food restriction indeed delays aging in humans and its effects are most dramatic when begun in adolescence as in rodents, how many people would want to risk growing into a miniature, thin-boned adult in order to purchase a bit more youth?

I attended a conference several years ago convened by the National Institute on Aging that included virtually every scientist in the United States who studied food restriction and aging. Surveying the room was pretty much like surveying America. There were a few lean folks like Walford (although he was not as lean as he once was), somewhat more were struggling with obesity, and the majority were just about our plump average. That may say it all.

I do know one other aging researcher who tried for years to restrict his own diet to a level (1,200 calories per day) that he thought was analogous to that of a food-restricted mouse. He told me that the worst part, as any less fanatical dieter can tell you, was that he always fantasized about food, a rather limited fantasy life as far as I'm concerned. When I asked him whether he felt this diet had been keeping him young, he honestly responded, "I don't know; I cheated too much to tell." He has since abandoned his food-restriction regime and now eats massive amounts of melatonin each day. (More on that, shortly.) Even the ideologically committed Biosphereans couldn't manage restriction that easily and ultimately had to place their food under lock and key to prevent pilfering.[116]

Exercise

Virtually every medical doctor and any newsstand health magazine or book with "antiaging" in its title recommends regular exercise as an essential ingredient for a long and healthy life. In one sense, the reason is obvious. Throughout human evolution, vigorous daily exercise has been essential for escaping predators and securing enough food to survive. Life-long exercise is as natural and normal as breathing. It has only been during the past few centuries at most that a sedentary life has become possible for large numbers of people. Physical inactivity, couch potato-dom, is a decided physiological abnormality, probably better thought of as exercise deficiency, a condition for which our bodies were not designed. It shouldn't be surprising, then, that exercise deficiency (especially combined with limitlessly available, high-calorie food) leads to shrunken, weakened muscles; an inability to exercise vigorously when required; increased blood cholesterol, glucose, insulin, and triglycerides; and, consequently, increased risks of developing atherosclerosis, high blood pressure, and diabetes.

Also, exercise seems to produce bodily changes opposite to those typical of aging. Specifically, as people age, they become less physically active, have less strength and endurance, and have a higher body fat percentage and less muscle. Exercise forces continuing high levels of physical activity, increases strength and endurance, decreases body fat, and increases muscle.

Exercise also increases metabolism. Individual muscles may increase their oxygen consumption as much as a hundredfold, and the overall metabolic rate during hard exercise may increase tenfold. If metabolism produces free radicals, higher metabolism will produce more damaging free radicals. What's more, body temperature also increases during exercise, which by all logic should accelerate the formation of browning products. If free radicals and browning products are fundamentally involved in aging, as virtually everyone now agrees, it stands to reason that exercise could actually accelerate aging, just as the rate-of-living guru, Raymond Pearl, predicted when he wrote his famous article, "Why Lazy People Live Longest" for the *Baltimore Sun* in 1927.

These views might be reconcilable if exercise didn't really increase total daily metabolism—something that many people believe. That is, people assume that increased metabolism *while exercising* is more than compensated for by *decreased metabolism* at other times. This, it turns out, is not true. Although highly conditioned athletes frequently have exceptionally low pulse rates while resting, pulse rate is not a good indicator of metabolism (only of heart efficiency). Vigorous exercise may even boost metabolism during the rest of the day (studies differ on this issue), which would increase overall metabolism even more.

So does exercise accelerate or delay aging? The simple answer is that there is no simple answer. We know a great deal about exercise and aging in rats; we know very much less but still a considerable amount about how it affects humans in industrialized societies. In nonindustrialized countries, exercise is not an issue. Getting through the average day provides all the exercise anyone could use. In rats and industrialized countries, though, there seems little doubt that exercise can (but doesn't always) increase longevity in a minor way. It doesn't delay aging.[117]

This isn't really the contradiction it seems. First, let's consider what the rats tell us, because with rats we can measure and manipulate exercise over a whole lifetime and measure precisely how it affects aging and survival. Whether rats live longer if they exercise depends on whether they exercise voluntarily (an exercise wheel is placed in the cage, and they use it at their whim) or by force (they are made to run on a motorized treadmill). I think of these as the business-executive versus boot-camp modes of exercise research.

Boot camp for rats, as for humans, can be beneficial or not, depending on the details. The difficulty is in knowing whether your rats are being worked or overworked, especially when they are elderly. Imagine trying to run a Marine boot camp with 60-year-old men. Some would thrive, no doubt; others would pitch face forward on day one. Forced exercise, then, can shorten life (like a Welsh coal miner, a rat *can* be worked to death), have no effect, or even extend life a bit. Generally, forced exercise is more likely to be beneficial when begun early in life but more likely to be detrimental when begun late. Voluntary exercise is much more predictable. It lengthens *average* rat life and tones the muscles and cardiovascular system, but it has no effect on how long the oldest rats in a group will live. That is, exercise seems to inhibit diseases of middle age but not to affect late-life ailments. This is how life can be lengthened without delaying aging.

A cautionary note about the relevance for humans of all exercise studies of laboratory animals is that, unlike humans and virtually all other animals in the real world, laboratory rodents, remember, have been selectively bred to thrive in the forced inactivity of their shoebox homes. Wild rats and mice, even if laboratory-reared in the same shoebox cage without previous exercise, can run many times as fast and many times as far as any laboratory animal, given the opportunity. Conceivably, exercise may affect the longevity of wild rodents very differently from the way it affects laboratory animals.

But one reason to suspect that humans and laboratory rats respond rather similarly to exercise is that humans also seem to have slightly longer lives, on average, if they exercise *in the proper way*, though aging remains unaffected. In other words, as in rats, there is no evidence that exercise has anything to do with living an extremely long life. The ranks of centenarians are not dominated by former Olympic swimmers or marathon runners—the reverse may even be true.

Also, remember that aging and longevity are not synonymous. Aging is the rate of generalized physiological deterioration. One way to measure generalized deterioration of at least cardiovascular and muscle function is to determine a person's ability to perform vigorous exercise on a treadmill, which measures aerobic capacity, or maximum oxygen use. Healthy sedentary—that is, exercise-deficient—people lose about 1 percent of their aerobic capacity per year beginning at 25 to 30 years of age. Regular vigorous exercisers, not surprisingly, have greater aerobic capacities than sedentary people at all ages; however, in vigorous exercisers as a group, aerobic capacity also declines by about the same 1 percent per year beginning at the same age.

Translating this into concrete performance terms, it is both heartening and depressing to browse the history of Olympic records. Unlike social and political history, athletic history allows us to justly claim we are better now than we were then. Improved training schedules, the use of videotape and sophisticated medical devices, knowledgeable trainers, better diets, and no doubt greater motivation (remember when Olympic athletes were careful not to become too obviously wealthy during these amateur years?) have increased performances by 20 to 90 percent, depending on the event. At the same time, certain things are depressingly the same. Virtually all records in events requiring strength, speed, or endurance are held now, as then, by 26- to 30-year-olds. World records in competitive running events categorized by age (no one is driven if not a world-class master athlete) notably decline by about the same 1 percent per year as maximum aerobic capacity in distances from one mile to 10,000 meters.

Exercise training, even begun relatively late in life, can increase strength and endurance. Does this mean it is rejuvenating? Let me put it another way. Sylvester Stallone is about my age and no doubt stronger than I am (although I might bet on me for endurance). Does this mean his muscles (or his body) are younger than mine? I hope not. I think that he is just closer to his strength potential, or his strength potential is greater than mine, or both. A useful way to think about exercise and other putative antiaging therapies is that everyone has a certain health potential, whether it be muscle or memory or immune-system potential. An appropriate diet or exercise program, even when begun later in life, may allow you to approach that potential more closely, but the potential itself wanes with age as inevitably as boxing champs are ultimately embarrassed by cocky whippersnappers. Only therapies that slow, stop, or reverse that declining potential are properly called antiaging therapies.

Of course, living closer to your aerobic potential will increase your ability to perform a wide range of activities, may retard some mid-life diseases, and certainly will enhance your quality of life—assuming that quality of life has anything to do with playing with the grandchildren, walking to the store, or continuing to live outside a nursing home until a ripe old age—even if it doesn't slow aging.

To those of us who like to work out maniacally, however, the life-lengthening effect even of proper exercise is disappointingly small, and maniacal exercise may not be the best way to go.

The best large human study of exercise and longevity has come from 30 years of tracking the fate of more than 17,000 men who graduated from Harvard University during the first half of this century. Now ranging in age from their mid-sixties to nineties, these men periodically respond to medical questionnaires on many aspects of their health and physical activities. Determining when the men die is no problem. Harvard, ever mindful of its financial endowment, tracks its alumni with all the care that hunters track big game.[118]

By the mid-1980s, this study had concluded that moderate physical activity, equivalent to a weekly regimen of climbing 250 to 450 flights of stairs, jogging three to five hours, or walking 20 to 35 miles, could purchase an extra one to two years of life relative to men who exercised either more or less. A later study of the same, but now older, Harvard alums found that only vigorous exercise (brisk walking, jogging, swimming laps, etc.) increased longevity even this much (or this little, depending on your point of view). Light, some might say pleasurable, activities such as gardening or golf had no effect, regardless of how often they were done. Of course, someone also had to point out that the one to two years gained were just about the amount of time you would spend actually doing the exercise over 40 or so years. Think of it as God offering you two extra years of life provided that you spend them jogging.

But you can apparently exercise too vigorously for optimal health. There is a depressing litany of aging runners who, like the jogging evangelist Jim Fixx (dead at 52), have dropped in their tracks at relatively young ages. This may be the cumulative effect of massive and prolonged free-radical production finally overwhelming those cellular antioxidant defenses. The physician Ken Cooper, whose 1968 best-seller *Aerobics* helped bring exercise obsession to public consciousness, now says that ultra-exercisers need a daily cocktail of antioxidant vitamins to combat the massive number of free radicals they produce. Whether there is evidence that this works for ultra-runners as well as for normal humans is a topic that we'll discuss shortly.

What we do know about the effects of exercise on free-radical defenses is complicated and inconsistent. In some parts of the body, defenses seem stimulated by exercise; in others, they are not affected; and in still others, they may be lowered. There certainly is no general enhancement, so why exercise doesn't accelerate aging is puzzling. At this point, we don't know. Perhaps it *does* accelerate some aspects of aging, but this is more than compensated for by other benefits. More likely, exercise affects our cellular chemistry in more subtle ways than we yet understand.

Before one is tempted to abandon the gym and jogging track because of this information, think about this: Exercise unquestionably improves mid-life health and vigor by a variety of measurements. The information that exists from long-term studies of exercise and longevity depends on people's own estimates of their activity level usually at one point in their lives rather than over many years. We don't know whether exercise begun very early in life and continued throughout has a larger effect. We understand most about how exercise affects the longevity of middle-aged men and much less about its effects on the elderly or women.

But clearly, exercise by itself is not the fountain of youth.

There is some intriguing new evidence—very preliminary and properly interpreted with caution—suggesting that moderate exercise *combined* with moderate food restriction could slow aging as much as full-blown, weep-on-your-knees food restriction. At least in rats.

Roger McCarter is an urbane, bespectacled muscle physiologist brimming with charm and civility who has done many of the rat experiments I've described. He is also a meticulous and thorough scientist, not given to self-advertisement and hyperbole about his own research. By chance, he happened to start some rats on a mildly restricted diet, eating only 10 percent less than *ad lib*. This is typically too little restriction to find an antiaging effect, but he also happened to give each rat the business-executive-treatment—an exercise wheel in its cage. To his astonishment, these rats have aged nearly as slowly as his previously most restricted animals. McCarter isn't ready to trumpet any huge breakthrough at this point. He wants to repeat and refine the work, as a properly cautious scientist should. But if this finding is confirmed—and even more tenuously, if it extends to humans—modestly dedicated folks might be able to manage a daily workout while eating 10 percent less in order to live another 40 to 50 years. I might even be willing to give it a go if the data come in before it's too late.

Antioxidants

If the production of oxygen free radicals—oxidants—contributes to aging in a major way, then it would seem inherently reasonable to assume that anything one can do to boost antioxidant defenses might allay aging. Yes, it might seem perfectly reasonable, and many people now consume megadoses of the holy antioxidant trinity—vitamins A, C, and E—under that assumption. Is there any evidence that antioxidant megadoses actually retard aging? The short answer is no. Might antioxidants be health-enhancing nevertheless? Might megadose vitamin supplementation be harmful in any way? These are other issues altogether.

Let's start with what we know for sure. Dozens of epidemiological studies agree that a diet rich in fruits and vegetables (both excellent sources of the antioxidant vitamins) and low in animal fat is correlated with modestly reduced incidences of cancer and heart disease, which, over a lifetime, should lead to small increases in average longevity. The nutritionists' grail for the past decade, at least, has been the discovery of the key ingredient or ingredients in fruits and vegetables responsible for this effect. Researchers have often leapt to the conclusion that the observed disease reduction must be due to the amount of antioxidant vitamins in fruit and vegetables. However, it isn't at all certain that this is the case. Any fruit or vegetable will contain at least 150 compounds, many of them antioxidants, in addition to vitamins A, C, and E. Any or all of these other compounds could be involved in reducing disease. Compounds that aren't antioxidants may also be involved, as may subtle lifestyle differences (things as grossly obvious as smoking have been controlled for in most of these studies) in people prone to eating lots of fruits and vegetables. Current knowledge doesn't allow us to sort these things out.

However, we also know that studies of healthy animals fed various antioxidants have been disappointing. About half the studies show marginal increases in average life, with little or no increase in maximum longevity. The other half find no effect whatsoever. Even some of the small increases might have been due to the reduced palatibility of the adulterated food or reduced appetite, that is, food restriction. Food consumption has usually not been measured.[119]

Needless to say, the nutritional requirements of humans differ dramatically from those of rats and mice. Rats would be no happier eating my steak than I would be eating their lab chow. Rats manufacture their own vitamin C, but we don't. They need a chemical called choline to prevent cirrhosis of the liver; we don't. So trying to generalize from rodents to humans about specific nutrients is a fool's errand. In this field, we know what we know from human studies.

In another sense, the lack of an antiaging (or even disease-retarding) response in feeding specific antioxidants to animals isn't so surprising. Knowing the chemistry—that a certain compound defuses free radicals in a test tube—means little about its effects in as complex a biological system as an animal's body. Cyanide, for instance, is a fine antioxidant, although its one rather unfortunate side effect, death, makes its medical usefulness limited. Living systems, unlike test-tube chemical reactions, thrive on a delicate equilibrium, something like a well-tuned engine. More of a good thing is not necessarily better. Engines require air, fuel, and fire in the cylinders, but no one who knows engines would suppose that the performance of a well-tuned engine would be enhanced by adding twice as much air to the mixture. Similarly, oxygen is vital for life, but too much oxygen, as we've seen, is fatal. Balance is the key. Remember this when you hear someone say, "What the hell: Even if it doesn't work, what harm can it do?"

Also, vitamins—which are defined as substances our bodies need in trace amounts but do not produce—like most biological chemicals and repertory actors, play many roles. Vitamin C, for instance, not only plays an important role as an antioxidant, it also helps the body produce normal collagen. Vitamin A plays roles in the proper development of bones and eye pigments critical for vision. The best amount for one function may be too much or too little for another.

Notice, too, that mainstream medicine never recommends that you include the smorgasbord of exotically named antioxidants I mentioned earlier—superoxide dismutase, glutathione peroxidase, catalase—in your diet. You may have wondered why. The reason is that these antioxidants are produced within your cells. They are proteins, which are disassembled like Legos by the enzymes in your stomach. Eating a gallon of superoxide dismutase would not increase its presence in your body (although these bogus health products are available at many health-food stores). The antioxidants you read about are *dietary* antioxidants, which your body doesn't manufacture. You have to eat them. Dietary antioxidants may not be as effective as internally produced antioxidants if for no other reason than that they may not get to the necessary location inside the cell. In molecular medicine, as in ballistic missiles, targeting is everything.

So if we all agree that eating a diet rich in fruits and vegetables is healthy, even if it doesn't retard aging per se, let's look at whether it makes potential sense to supplement your existing diet with specific antioxidant vitamins.

Vitamin C

Linus Pauling won two Nobel Prizes, and if things had broken differently, he might have won a third. When double-Nobelists speak, the world listens. Pauling's chemistry prize in 1954 merely confirmed what many chemists already knew—that he was probably the most brilliant chemist of his generation. The other prize, the one that shot him from the obscurity of the laboratory to celebrity status, was the Nobel Peace Prize in 1962 for his efforts against nuclear proliferation and weapons testing, and against war generally as a means of solving international conflict. A few years later, these efforts made him an intellectual godfather to the antiwar movement of the 1960s.

When such a widely admired chemist and well-known public figure speculated in print in 1970 that massive doses of vitamin C could prevent and treat the common cold, drugstore supplies of it were quickly sold out. The vitamin supplement biz flourished and today rakes in $3.5 billion per year, largely thanks to Linus Pauling.

As time passed, Pauling also became convinced that vitamin C, because of its antioxidant properties, could also be effective in preventing and treating cancer and heart diseases. He founded a research institute to investigate vitamin C's effects on health. It didn't hurt his credibility that he lived into his nineties and that he charged ahead with his research, even though the medical establishment reflexively pooh-poohed his claims and supported its skepticism with some rather shoddy research.[120] Everyone loves Don Quixote.

But brilliant scientists are not necessarily brilliant about everything. J. B. S. Haldane, a scientific hero of mine, admired Stalin for much of his life, and a lot of Depression-era German physicists thought that this Hitler thing was really quite exaggerated by the foreign press. Pauling was nearly 70 when his preoccupation with vitamin C began, and a lot of us wondered whether this was just one of those unaccountable late-life obsessions, like Ruskin's with Rose LaTouche.

Vitamin C is found only in fruit and vegetables, particularly in green peppers, broccoli, citrus fruit, strawberries, melons, tomatoes, raw cabbage, spinach, and other leafy greens. A deficiency of vitamin C causes scurvy, that grisly bane of lost and becalmed sailors. When an eighteenth-century British physician discovered that eating limes could prevent scurvy, British sailors became forever after "limeys." Scurvy is prevented by minuscule quantities of vitamin C, about 10 milligrams per day, contained in less than an ounce of cooked broccoli or orange juice.

The current U.S. government RDA (Recommended Daily Allowance) for vitamin C of 60 milligrams is designed to prevent scurvy even if no vitamin C is consumed for four to six weeks. The RDA is easily met with a normal balanced diet. By contrast, the PRDD (Pauling Recommended Daily Dose) was 3,000 to 12,000 (!) milligrams, 200 times the RDA. He felt that this dose might give people an extra 12 to 18 years of life, and the pharmaceutical company Hoffmann-LaRoche loved him for it. A cow would have trouble eating enough broccoli to get the PRDD. You'd have to take supplements. Hoffmann-LaRoche is vitamin C-supplement wholesaler to the world.

Given the extreme disparity between the RDA and the PRDD, is there evidence that the Pauling dose, or other megadoses, might be harmful, let alone life-extending? The public perception is that megadoses of Pauling's C, because it is water-soluble, are harmless. Excess is simply washed away in urine.

But professional nutritionists, at least those not employed by vitamin-supplement manufacturers or with a vested intellectual interest in specific vitamins, are not enthusiastic about massive vitamin supplements for some good reasons. First, they can interfere with other health-related laboratory tests. For instance, high-dose vitamin C supplements interfere with some common and important measurements of laboratory blood tests.[121] Supplements also interfere with detection of blood in urine and stool specimens. Second, supplements can alter the absorption of other nutrients in complex, frequently unknown ways. For instance, vitamin C decreases copper and increases certain types of iron absorption. Third, supplements may not contain the amounts of product claimed on the label or may be contaminated.[122] Because supplements are classified as food rather than as drugs, the U.S. Food and Drug Administration has no regulatory control over them.

Aside from these issues, massive doses of vitamin C are probably not harmful[123] (although little evidence exists one way or the other for long-term studies) except to two groups of people. One group consists of people, maybe as many as one-sixth of Americans, who are genetically prone to the formation of certain kinds of disablingly painful kidney stones. For these people, high doses of vitamin C may increase the chances of developing stones. Also, people with a genetic propensity to store exceptionally high levels of iron in their bodies, as many as 12 percent of Americans, might find megadoses of vitamin C dangerous. In the presence of excess iron, vitamin C paradoxically turns from an antioxidant into an oxidant, or free-radical producer, and can potentially lead to some of the problems—increased atherosclerosis, cancer, cataracts—that seem to be associated with free-radical damage.[124]

Given these cautions, is there any compelling evidence that vitamin C does anything beyond keeping supplement makers solvent? It doesn't stop, slow, or reverse aging, of course, but some epidemiological evidence suggests that supplements are weakly associated with a decrease in some types of cancer (digestive tract, lung, uterus), but not others (breast and prostate). Again, studies of diets high in fruits and vegetables show a more consistent and stronger effect on reducing cancer incidence.[125] There is no similar evidence on vitamin C supplementation for heart disease.

Nor will there be any better information on vitamin C alone in the near future. Although three large clinical trials with vitamin C supplements are currently under way—one in the United States, one in the United Kingdom, and one in Continental Europe—all the studies combine vitamin C with several other vitamins. We may know more about *combinations* of vitamin supplements before too long.

If you insist on taking vitamin C supplements instead of just eating more fruit and vegetables despite this weak evidence, how much should you take? Again, we have limited information, but the most thorough investigation comes from a recent study of seven subjects who were monitored so closely they had to spend more than four months in a hospital being poked, bled, and fed. For this small group of 20- to 26-year-old men, the "best" daily dose—that is, the dose that led to the highest blood levels without production of potentially harmful byproducts—was 200 milligrams, more than four times the RDA but less than one-tenth the PRDD. It is important to note that these were all young men who had been thoroughly screened for diseases associated with the more toxic effects of vitamin C. No similarly reliable information exists for women or for middle-aged and elderly people of either sex. Let me also reemphasize that 200 milligrams of vitamin C will automatically be consumed by those following current FDA dietary guidelines of at least five servings of fruit and vegetables per day.

Consumption (Recommended and Otherwise) of Antioxidant Vitamins

Supplements	RDA*	Amounts consumed per day by some people
Vitamin C (ascorbic acid)	60 mg	1,000–12,000 mg
Beta-carotene	5–6 mg	20–50 mg**
Vitamin E (alpha-tocopherol)	8–12 mg	50–200+ mg

*RDA for adults.
**Amounts used in the various clinical trials.

Vitamin A

Vitamin A, or retinol, occurs naturally only in animal products such as liver, butter, egg yolks, whole milk, and particularly fish oils. Skim milk and margarine are typically spiked (that is, fortified) with vitamin A. However, the body can convert some plant chemicals called carotenoids, especially beta-carotene, into vitamin A. Dark-green leafy vegetables as well as yellow and orange fruit and vegetables are rich in carotenoids, including beta-carotene.

As is the case with all vitamins, if you eat too little vitamin A, you will run into serious problems. In fact, vitamin A deficiency is the leading cause of blindness in children worldwide. On the other hand, vitamin A is stored in the liver, and you can poison yourself with too much. Early Arctic explorers may have owed some of their numerous fatalities to dietary overreliance on polar bear livers, which fairly drip with vitamin A. Signs of vitamin A poisoning include aching bones, severe headaches, and what is medically called hyperexcitability. None of the above seems conducive to successful Arctic exploration.

For years, we thought we had the solution to the overdose problem, though. Because beta-carotene needs to be converted to vitamin A, and because there are physiological controls over the rate of conversion, consuming beta-carotene instead of vitamin A itself was thought to insure against vitamin A poisoning. The worst you seemed to be able to do with beta-carotene was turn your skin the color of carrots.

At the same time, a large, fairly consistent body of observational epidemiology pointed to reduced cancer, especially lung and stomach cancer, among people with diets or blood that was chockablock with beta-carotene (that is, people who eat lots of fruit and vegetables). So persuasive was the epidemiological evidence that in the mid-1980s, health authorities around the world began at least six clinical trials—that is, properly controlled experiments—on the health effects of beta-carotene and other antioxidant vitamin supplements. In the last couple of years, results from those trials have begun piling up like rugby players in a scrum. The news has been less than encouraging.

It is now pretty obvious that beta-carotene is not the gold nugget hidden beneath the pile of the fruit and vegetables. First came the now infamous Finnish smokers study. Nearly 30,000 men in their fifties who had been smoking for 35 years, on average—high-wire lung-cancer risk takers—were given a daily placebo, or 20 milligrams of beta-carotene, or 50 milligrams of vitamin E, or both, and followed for six years.[126] Blood levels of beta-carotene increased more than tenfold among the supplement users, yet their death rate *increased* by 8 percent, mostly because of more lung cancer. The increase in lung cancer was thought to be a statistical fluke, like happening to flip heads on five coins in a row. Surely beta-carotene couldn't lead to more cancer. Yet two years later, a second large trial found the same thing. This experiment, the Beta-Carotene and Retinol Efficacy Trial (CARET), fed a combination of beta-carotene and vitamin A to American smokers, former smokers, and asbestos workers (who are at especially high risk of lung cancer) of both sexes, but after four years the experiment was abruptly cancelled when researchers found that those receiving the supplement had a 28 percent higher risk of developing lung cancer and a 26 percent higher risk of death from cardiovascular disease than the placebo takers. A third large experiment found no increase in cancer or heart attacks among healthy, supplement-taking male doctors. Tracing the fate of these men for 12 years, researchers found virtually identical death and disease rates among the placebo and the supplement takers.

These results have affected other medical experiments. Smokers in a trial of beta-carotene's effect on eye disease were advised to stop taking the supplement or else sign a consent form acknowledging the risk. Another trial, the Women's Health Study, removed beta-carotene from its supplement protocol.

Needless to say, vitamin-supplement makers and takers were quick to attack these studies. They had given too much (or too little) of the supplement. They were not giving the right *combination* of antioxidants. The subjects were unhealthy to begin with. What do you expect of smokers' health anyway? But the fact remains that in large strictly controlled and analyzed experiments, beta-carotene supplements either increased or had no effect on the risk of death and disease. To date, overall consumption of vitamin supplements appears unaffected, though. Evidence is not always decisive when it comes to strongly held beliefs.

Vitamin E

Unlike A and C, the role of vitamin E in normal adults has been difficult to determine precisely, although it probably has a role in antioxidant protection of cellular membranes. About the only people found to have vitamin E deficiency are those with a defective ability to absorb dietary fat. Some beneficial effects of supplemental vitamin E have been reported in premature babies and infants with particular medical problems.

In spite of the fact that it is fat-soluble, vitamin E doesn't seem to be stored in any specific organ. The only potentially harmful effect of excess vitamin E consumption is that it may interfere with one's ability to absorb other vitamins (specifically A and D). Vitamin E is found in vegetables and vegetable oils, nuts, and green leafy vegetables. Corn and soybeans are particularly rich sources.

211

In a test tube or dish of cultured cells, vitamin E seems to have precisely those antioxidant activities which should combat heart disease and cancer. Epidemiological studies have been variable—the major trend being that diets high in vitamin E show more consistently beneficial effects than supplements high in vitamin A. The same clinical trial of Finnish smokers that found elevated lung cancer and overall deaths among those receiving beta-carotene supplements found no similar increase of death or disease among takers of 50 milligrams per day of vitamin E. They found no decrease in death or disease, either, although a follow-up showed a vanishingly small decreased risk of developing angina, a sign that too little oxygen is getting to the heart. There are at least four continuing trials of the effects of vitamin E on heart disease, and those results should clear up what looks at present to be a not particularly promising avenue of disease prevention.

If lots of fruit and vegetables seems to be related to reduced cancer and heart disease, wouldn't it make sense to take a smorgasbord of "phytochemicals"—that is, specific extracts from plants? Maybe, maybe not. We still don't know what the key ingredients in plants are, or even if they exist—although there is some reason for optimism on that point. Gobbling random plant chemicals probably isn't a good idea, because we already know that plants contain lots of compounds that in large amounts cause cancer and birth defects. As I said earlier, the biochemist Bruce Ames has calculated that 99.99 percent of the carcinogens we eat are natural products. Eating plants generally is good; eating specific chemical components of plants may not be. On this one, it would be most prudent to let experiments separate the poisons from the potions for us.

The take-home message about antioxidant vitamins is depressingly trite. Your mother (and even more amazingly, the FDA) seem to be right. Eat lots of fruit and vegetables and lower your fat intake, and you can't go wrong. It won't slow aging, but it will reduce your risk of specific important mid- and late-life diseases. Too little of virtually any vitamin, including these antioxidants, is not good for you. On the other hand, regular use of vitamin pills containing vastly more than the recommended daily allowances is not likely to compensate for a poor diet or other bad health habits, and may even do you harm. If there is a silver antiaging bullet out there, the antioxidant vitamins are not it.

Melatonin

Might that bullet be melatonin? In terms of hope and hype, melatonin was the vitamin C, or maybe the Lourdes, of 1995. A flurry of new books, one even written by a respected scientist, claimed that melatonin could do everything from prevent or cure heart disease, Alzheimer's disease, and cancer, to rev up your sex life and fight AIDS, epilepsy, and Parkinson's disease. It was also likely to allow you to sleep better, lose weight, and, oh yes, delay aging. Melatonin, apparently, does everything except the laundry. Moreover, it was a newcomer. Most panaceas for aging have been making their claims for years. Melatonin was a late-bloomer, but is making up for lost time fast.

The idea that melatonin supplements might slow aging is consistent with what you might think of as the Macerated Testicle Principle of rejuvenation. Remember that Charles-Edouard Brown-Séquard injected himself with animal-testicle extracts because he felt his normal sex drive waning. The rationale is that if something declines with age, you just need to top it up periodically in order to stop, or at least slow, aging. It sounds ridiculous, and by and large it is. But just as paranoiacs can have real enemies, overhyped hormones might have real benefits. So what do we really know about melatonin, and what do we wish we knew?

In the dead center of your brain is a pea-size gland called the pineal, which is the source of most of the melatonin produced in your body. The pineal secretes five to 10 times as much melatonin at night as during the day. This potentially allows the body to count days and, by keeping rhythm with the changing length of the day (actually, night) throughout the year, helps animals adjust their activities, such as breeding, to the changing seasons. As we age, our pineals secrete less and less melatonin at night.

There's no doubt that melatonin is intriguing. In laboratory rodents, extra melatonin seems to protect against the free-radical effects of strong carcinogens and lowers cholesterol, among other things that seem to be consistent with increased protection against the diseases of aging. There are other effects you less often hear about. Melatonin in high doses inhibits ovulation, for instance, and a melatonin-based contraceptive will probably be on the market before long. In humans, reasonable evidence also exists that melatonin is a pretty good sleeping pill and can mitigate jet lag. What about increasing longevity?

Here the story goes a bit out of focus. As I said before, people rarely claim antiaging activities of substances for which they don't have a patent or a vested intellectual interest. One research group in particular has reported amazing effects of melatonin on mouse longevity. Just by chance, these same people also have writtten the book making the most wildly extravagant claims about the antiaging effects of melatonin.[127]

They found for instance that spiking the drinking water of mice with melatonin extended their lives by about 20 percent. Furthermore, and this is the really astonishing part, they found that transplanting the pineal from a young mouse into an old mouse extends the old mouse's life by about the same 20 percent. What makes this experiment so astonishing, as pointed out by another melatonin researcher, is that they found this effect in strains of laboratory mice that, because of a genetic defect, cannot manufacture melatonin in the first place. Perhaps their pineals secrete Gerovital H3 or essence of goat testicles.[128] In reality, all the laboratory mice commonly used for aging research have the same genetic defect that prevents them from manufacturing melatonin. Yet these mice are not short-lived by mouse standards. They sleep, have sex, and become sexually mature, just like other mice. Considering how vital melatonin has been claimed to be, these mice are doing pretty well. Also, melatonin levels peak in humans at about five years of age—before we can properly be said to be aging at all.

Does this mean that all the claims for the health effects of melatonin are bunk? No. It does mean that most such claims are theories, far from established facts that they are often presented as being, at least for humans. Yes, indeed. One decent book on the topic exists,[129] but even reading it you have to pay close attention to whether the effects described were observed in cells in a dish, in rats, or in people. Little experimental work exists on melatonin's effects in people. Observational studies, such as finding that sick people have less melatonin than healthy people or old people have less than youngsters, hardly establish that melatonin is responsible for their sickness. Your stuffy nose does not cause your cold.

Well, what can you lose? Isn't this another case of "What the hell: Even if it doesn't work, what harm can it do"? After all, if it is sold over the counter, it must be harmless, right? Tell it to the French. In France, and a number of other European countries, melatonin is not an over-the-counter drug. You get it by prescription only. Why? Side effects are theoretically possible (although not established to occur) in:

1. people taking steroid drugs, such as cortisone or dexamethasone;
2. pregnant women;
3. women wanting to get pregnant;
4. nursing mothers;
5. the mentally ill;
6. people with allergies or autoimmune diseases;
7. people with immune-system cancer.[130]

If nothing else, the recent attention attracted by melatonin has stimulated research by more than the handful of laboratories that were previously investigating it. Stay tuned. The false versus the real hope offered by melatonin should be determined in the next few years.

DHEA

DHEA, or dehydroepiandrosterone, is a hormone without a function, as the endocrinologist Jim Nelson puts it. Here you have the most abundant steroid in the body (some of the others are testosterone, estrogen, progesterone, and the stress hormone cortisol) and no one

215

seems to know what it does. It can be turned into testosterone and estrogen by the body, and stress suppresses it.

For more than a decade, this hormone has attracted the attention of aging researchers because, unlike other steroids, its level in the blood declines rapidly after puberty, and when injected into or fed to laboratory rodents, it appears to suppress tumor formation and guard against obesity, diabetes, and some immune and heart diseases. Its effect in humans is not clear, although it is widely used in Europe to treat discomfort associated with gout and menopause. DHEA is not an approved drug in the United States. It can be purchased over the counter, although there are no controls over the content or purity of commercial supplies. A recent study of capsules labeled as containing 500 milligrams of DHEA plus "natural precursors" found less than 15 milligrams of the hormone per capsule. Is this just the Macerated Testicle Principle once again, or is there more to it?

Alas, we don't really know at present, but there are clearly reasons besides drug purity to be cautious about gobbling this product. The most dramatic effects of DHEA have been observed in laboratory mice strains, which produce vanishingly small amounts of the hormone on their own and have special genetic predispositions to cancer and immune-system diseases. These mice, in fact, outlive untreated mice, but it isn't clear that this effect isn't due to appetite suppression, and thus unintentional food restriction. Long-term administration of large amounts of DHEA to rats has led to liver cancer.

There is no way to relate the rodent studies to humans. DHEA and the other steroids occur in minuscule doses in rodents compared with humans, and steroids interact with one another in complicated, poorly understood ways. That some steroids are involved with some human cancers is not controversial. The few human trials that have been done, which used very modest DHEA doses compared with those given in the rodent studies, have shown some improvement in immune response, muscle strength, and sleep patterns among the elderly.

DHEA deserves wider attention among researchers. Too much of the current work is done by True Believers, not properly agnostic scientists. Although almost all of the effects of DHEA on humans are unknown, there is enough smoke to warrant more investigation, though not nearly enough to conclude that there is a fire.[131]

Deprenyl

Among the most successful drugs for treating Parkinson's disease, most often noticed as extreme trembling of limbs and a "masklike" face in the elderly, is deprenyl (also known as Eldepryl®, selegiline hydrochloride, or L-deprenyl). In at least three separate experiments, treatment of elderly male rats with deprenyl at about twice the therapeutic human dose led to substantially increased average and maximum longevity (life extension ranged from about 10 percent to nearly 40 percent). One of the three studies also reported rejuvenation of sexual activity and improved learning in the rats. A fourth study found that deprenyl did not reduce appetite at least in young rats, so inadvertent food restriction is not likely to explain these results. Significantly, the same study found that after three weeks of deprenyl, the brain region typically affected by Parkinson's disease showed increased levels of cellular antioxidants.[132]

In male rats, at least, deprenyl appears to retard aging through some mechanism that isn't related to food restriction. That mechanism may be enhancement of free-radical defenses. One enzyme known to be specifically inhibited by deprenyl is found in mitochondria, the site of by far the greatest free-radical production in the cell.

As intriguing as these experiments are, they shouldn't be overinterpreted. The same antiaging effect has yet to be shown in young rats, female rats, or mice of any age, much less in humans. Deprenyl chiefly inhibits only one of two forms of this particular enzyme (specifically, that form is called MAO-B), which is much less common in rats than in humans. Who knows the significance of that? The pharmaceutical company that markets deprenyl for the treatment of Parkinson's disease warns adamantly about the potential side effects of higher than recommended doses. Remember, rats were given about twice the recommended human dose. Finally, a separate study failed to find the same enhancement in antioxidant levels as that found in the brain region affected by Parkinson's disease in either blood or cerebrospinal fluid.

As with melatonin and DHEA, the deprenyl story may represent a dead end or perhaps the shadow of a ghost of a breakthrough. I'm not particularly sanguine about the impact of just restoring youthful levels of naturally occurring hormones. The Macerated Testicle Principle is not something I believe in. Deprenyl sounds to me much more promising. However, strong conclusions one way or another about the human antiaging effects of any of these compounds are premature. Existing information is sufficiently provocative to warrant vigorous and extensive follow-up. And if research money holds out, vigorous and extensive follow-up will surely come about. I tend to think of these putative remedies not as far-fetched, but as short-fetched. That is, it won't be long before their antiaging potential, or lack of it, is firmly nailed down. Therapies for aging that are farther down the research pipeline, but to me seem more likely to lead to real changes in the aging rate, deserve a short section all their own.

Expectations

For someone from a family of eminent scientists, the writer Aldous Huxley hated and feared the notion of scientific progress—especially with regard to aging—with exceptional vigor. In one of his books,[133] science achieves a 200-year-old man, but only by turning him into a scratching, chattering, chimpanzoid monstrosity. In another,[134] the future produces 60-year-old men who remain untouched by the stigmata of aging, with all the vigor and alas values that they had at 17. In return, they sacrifice leisure. There is no time to sit and think. Whether either of these scenarios seems as horrifying to us as they did to Huxley, what he imagined now seems as anachronistic as those magazine articles from the 1950s projecting that we would by now have entirely self-cleaning homes and be riding private rocket ships to work.

Predicting the future, in other words, is hazardous. Nevertheless,

I feel obliged to risk a few guesses about what seem to me to be the most likely prospects for real aging remedies to be developed in the foreseeable future. I won't discuss cures or treatments for individual diseases of aging, although progress will be made on this front as well. I'll confine my speculation to the treatment of aging itself.

It's not just a guess that within a few years the entire human genome will be mapped and sequenced. Currently projected to be completed early in the next century, the Human Genome Project has consistently progressed more rapidly than projections. Using these maps and sequences, gene therapy, in which defective genes are destroyed or replaced with normal genes, will become commonplace. We are already routinely transplanting human genes into bacteria, pigs, and mice and fiddling with regulating the action of individual genes in fruit flies. Currently, more than 100 clinical trials of gene-based therapy are under way. Although none of these well-publicized therapies has yet been successful, it is only a matter of time (and probably not a long time) before effective procedures are developed. It is in our increasing control over our genes and the deciphering of what certain gene products do that holds the best hope for retarding aging in the near future. Few of us think that controlling one or a few human genes will stop aging, but a measurable slowing seems within the realm of possibility. A few avenues of genetic research appear particularly promising.

We are closing in on understanding appetite and obesity. In 1994, a defective mouse gene causing obesity was identified and its human analogue located. Administration of the normal gene product, called leptin, to genetically obese mice decreased the amount they ate, increased their energy expenditure, and reduced their weight. Think of it as genetically induced dieting. The amount of the analogous gene product in humans is closely linked to the amount of body fat.[135] At least four other mouse genes that when defective cause obesity have also been identified. Humans have analogues of all these genes, but obese people, unlike obese mice, have not been found to be defective in these genes. But everyone agrees that we know vastly more about appetite and its control than ever before, and we are learning at an accelerated rate. At least a dozen new products helping to reduce appetite are in the research pipeline, some nearing approval, some still in clinical trials. No one is ready to claim victory over obesity yet. But significant steps are being taken.[136] The game may at least be in its late innings.

If—and it's a big if—food restriction delays aging in humans as it does in rats and mice, then manipulation of this and related genes might allow us to modify eating habits upon demand. Gene power instead of willpower. In principle, modified eating habits would afford us the opportunity to choose how quickly to reach puberty, when to have children, and how quickly to age—within limits.

Even if the antiaging effect of food restriction turns out to be a physiological idiosyncrasy of laboratory rodents, the possibility for genetic modulation of aging will by no means be exhausted. Other types of genetic therapy are at least as plausible. Fruit flies genetically engineered to produce additional amounts of two different antioxidants within their cells have already been shown to outlive nonengineered flies by about 25 percent.[137] It's a long step from fruit flies to humans, but who knows how general such phenomena might be? Studies are already under way to determine whether similarly engineered mice will be similarly long-lived. We should have the answer within the next four years, and if those results are promising, monkey studies will no doubt follow close behind, with human trials in the observable distance.

Greater production of existing gene products, such as the cellular antioxidants, is not the only other genetic engineering approach to aging, either. What about gene products from different species?

As previously noted, animals typically studied by the biomedical community are relative failures when it comes to aging. That is, worms, fruit flies, rats, and mice are all far less successful at combating such damaging aging processes as oxidation and browning than are humans. The utility of these species is that precisely because they are short-lived, multiple generations may be studied over the lifetime of a single research grant. We have learned, and will continue to learn, an enormous amount about fundamental biology from studying these species. However, progress in learning how to cope with the cellular ravages of aging might be better served by the study of animals that nature has made better than humans at combating these ravages. Nature, as someone put it, is smarter than we are. Because of their unique combination of high metabolism and long life, birds, for instance, may be exposed to five times as much oxygen per cell as humans during a lifetime. How do they manage to mitigate the damage? The same is true of browning, which should be a much more serious problem for birds, with their high blood sugar and body temperature, than for humans.

But given bird longevity, it apparently isn't. What protections have they evolved that we haven't? Elephants and whales have far more cells at risk of turning cancerous than do humans, yet they survive about as long as humans. Thus, they must have specially effective mechanisms for preventing cells from turning cancerous. What are they?

These questions used to be interesting in an abstract way—like wondering how many angels could dance on the head of a pin. However, methods are being developed that will allow us to begin counting those angels. Specifically, thanks to our ability to grow in laboratory dishes cells from a wide range of animals, and our increasing ability to identify and isolate genes associated with specific phenomena, we should be in a position before too many more years to begin putting bird antibrowning or antioxidation genes into mice, for instance, to try to duplicate the cellular damage resistance of birds. If our current theories about aging are accurate, this should represent a big step toward the 10-year mouse, and perhaps the 150-year human, life span.

After centuries of hokum and false hope, aging finally may be ready to yield to scientific manipulation. It is an exciting time to be alive and an especially exciting time to be a gerontologist.

Notes

1. The Paradox of Aging

1. W. J. Thoms, *Human Longevity: Its Facts and Fictions* (1873).

2. "161 Years Old and Going Strong," *Life*, September 16, 1966, 121–127.

3. L. Badash, "The Age-of-the-Earth Debate" *Scientific American*, August 1989, 90–96.

4. P. Medawar, *An Unsolved Problem of Biology* (London: H. K. Lewis & Co., 1952).

5. C. E. Finch, M. C. Pike, and M. Witten, "Slow Mortality Rate Accelerations during Aging in Some Animals Approximate that of Humans," *Science*, vol. 249 (1990), 902–905. The original data on Australian prisoners of war were compiled by R. A. Bergman in "Who is Old? Death Rate in a Japanese Concentration Camp," *Journal of Gerontology* 3, (1948): 14–20, and the original mortality-rate analysis was by H. B. Jones ("The Relation of Human Health to Age, Place, and Time," in *Handbook of Aging and the Individual*, ed. J. E. Birren (Chicago: University of Chicago Press, 1959), 336–363.

6. C. P. Lyman, R. C. O'Brien, G. C. Greene, and E. D. Papafrangos, "Hibernation and Longevity in the Turkish Hamster, *Mesocricetus brandti*," *Science*, vol. 212 (1981), 668–670.

7. S. J. Olshansky, B. A. Carnes, and C. Cassel, "In Search of Methusaleh: Estimating the Upper Limits to Human Longevity." *Science*, vol. 250 (1990), 634–639.

2. Age Inflation and the Limits of Life

8. V. Kannisto, *Development of Oldest-Old Mortality, 1950–1990: Evidence from 28 Developed Countries* (Odense, Denmark: Odense University Press, 1994).

9. A. Leaf, "Long-Lived Populations (Extreme Old Age)," in *Principles of Geriatric Medicine and Gerontology*, 2nd ed., ed. W. R. Hazzard, R. Andres, E. L. Bierman, and J. P. Blass (New York: McGraw-Hill, 1990).

10. Here are a few books on this topic: R. Taylor, *Long Life and Happiness* (1944); J. I. Rodale, *The Healthy Hunzas* (1948); J. M. Hoffman, *Hunza Secrets of the World's Healthiest and Oldest Living People* (1973).

11. R. McCarrison, *Nutrition and National Health* (London: Faber & Faber, 1944).

12. R. J. Myers, "Analysis of Mortality in the Soviet Union according to 1949–1959 Life Tables," *Transactions of the Society of Actuaries*, vol. 16 (1965), 309–317.

13. D. Davies, *The Centenarians of the Andes* (New York: Anchor Press/Doubleday, 1975).

14. Leaf, "Long-Lived Populations."

15. R. B. Mazess, and S. H. Forman, "Longevity and Age Exaggeration in Vilcabamba," *Journal of Gerontology* 34 (1979): 94–98.

16. S. J. Olshansky, "Introduction: New Developments in Mortality, in Symposium on Mortality Crossovers and Selective Survival in Human and Nonhuman Populations," *The Gerontologist* 35 (1995): 583–587.

3. Has Aging Changed over Time?

17. K. M. Weiss, "Demographic Models for Anthropology," *Memoirs of the Society of American Archaeology*, no. 27.

18. J. J. Angel, "Paleoecology, Paleodemography, and Health," in *Population, Ecology, and Social Evolution*, ed. S. Polgar (The Hague: Mouton, 1975), 167–190.

19. Wonderfully informative and readable accounts of the history of disease are W. H. McNeill's *Plagues and Peoples* (New York: Doubleday, 1976) and A. W. Crosby's *Ecological Imperialism: The Biological Expansion of Europe, 900–1900* (Cambridge: Cambridge University Press, 1986).

20. L. Heligman, N. Chen, and O. Babakol, "Shifts in the Structure of Population and Deaths in Less Developed Regions," in *The Epidemiological Transition*, ed. J. N. Gribble and S. H. Preston (Washington, D.C.: National Academy Press, 1993).

21. E. S. Deevey, Jr., "The Probability of Death," *Scientific American*, April 1950.

22. F. D. Zeman, "Old Age in Ancient Egypt," *Journal of the Mount Sinai Hospital* 8 (1942): 1161–1165.

23. E. Trinkaus and D. D. Thompson, "Femoral Diaphyseal Histomorphometric Age Determinations for the Shanidar 3, 4, 5, and 6 Neandertals and Neandertal Longevity," *American Journal of Physical Anthropology* 72 (1987): 123–129.

4. Is Aging Genetic?

24. A. G. Motulsky and J. D. Brunzell, "The Genetics of Coronary Atherosclerosis," in *The Genetic Basis of Common Diseases*, ed. R. A. King, J. I. Rotter, and A. G. Motulsky (Oxford: Oxford University Press, 1992).

25. Discussion of heart disease in Papua New Guinea and how changing social conditions have changed its frequency may be found in P. F. Sinnett et al., "Social Change and the Emergence of Degenerative Cardiovascular Disease in Papua New Guinea," in *Human Biology in Papua New Guinea*, ed. R. D. Attenborough and M. P. Alpers (Oxford: Clarendon Press, 1992), 373–386.

26. F. Schachter et al., "Genetic Associations with Human Longevity at the APOE and ACE Loci," *Nature Genetics*, vol. 6 (1994), 29–32.

27. A. G. Bell, *The Duration of Life and Conditions Associated with Longevity*, A Study of the Hyde Genealogy (Washington, D.C.: Genealogical Record Office, 1918).

28. M. McGue, J. W. Vaupel, N. Holm, and B. Harvald, "Longevity Is Moderately Heritable in a Sample of Danish Twins Born 1870–1880," *Journal of Gerontology* 48 (1993): B237–B244.

29. For information on mice, see R. Gelman et al., "Murine Chromosomal Regions Correlated with Longevity," *Genetics* 118 (1988): 693–704; for a nematode worm, see T. E. Johnson and W. B. Wood, "Genetic Analysis of Lifespan in *Caenorhabditis elegans*," *Proceedings of the National Academy of Science, USA*, vol. 79 (1982), 603.

30. These recent findings are summarized in G. M. Martin et al., "Genetic Analysis of Ageing: Role of Oxidative Damage and Environmental Stresses," *Nature Genetics* 13 (1996): 25–34.

31. United Nations, *Demographic Yearbook, 1988* (New York: United Nations, Dept. of International Economics and Social Affairs, Statistical Office, 1990).

32. D. W. E. Smith, *Human Longevity* (Oxford: Oxford University Press, 1993).

33. For information about how sex hormones affect the immune system, see

C. J. Grossman, "Interactions between the Gonadal Steroids and the Immune System," *Science*, vol. 227 (1985), 257–261. Information on the sex difference in disease susceptibility is in W. R. Hazzard, "The Sex Differential in Longevity," in *Principles of Geriatric Medicine and Gerontology*, ed. W. R. Hazzard et al. (New York: McGraw-Hill, 1994), 37–47. Also, information on sexual similarity in longevity among mice can be found in D. E. Harrison and J. R. Archer, "Physiological Assays for Biological Age in Mice: Relationship of Collagen, Renal Function, and Longevity," *Experimental Aging Research* 9 (1983): 245–251.

5. Why Does Aging Happen?

34. A. Comfort, *The Biology of Senescence* (Edinburgh and London: Churchill Livingstone, 1959, 1972, and 1979); C. E. Finch, *Longevity, Senescence, and the Genome* (Chicago: University of Chicago Press, 1990).

35. Z. A. Medvedev, "An Attempt at a Rational Classification of Theories of Ageing," *Biological Reviews* 65 (1990): 375–398.

36. Good books that discuss individual versus group selection are R. Dawkins's *The Selfish Gene, New Edition* (Oxford: Oxford University Press, 1989) and the more technical work of G. C. Williams, *Adaptation and Natural Selection* (Princeton: Princeton University Press, 1966).

37. This is an important point to make, because virtually all research gerontologists now agree that what has been called cellular aging is not actually aging itself. At the same time, researchers from other fields assume that anything that takes place at the cellular level must be the fundamental process. At least one new book on aging considers nothing but so-called cellular aging, and several textbooks currently used in courses in which I've lectured at major medical schools fail to mention any aspect of aging except at this erroneous cellular level. A technical review of the history and details of the Hayflick Limit phenomenon can be found in T. H. Norwood and J. R. Smith, "The Cultured Fibroblast-like Cell as a Model for the Study of Aging," in *Handbook of the Biology of Aging*, 2nd ed., ed. C. E. Finch and E. L. Schneider (New York: Van Nostrand, 1985), 291–321. A nontechnical firsthand account of the discovery is in L. Hayflick's *How and Why We Age* (New York: Ballantine, 1994).

38. Declining muscle strength and the effect of exercise is discussed in Chapter 10.

39. E. Chang and C. B. Harley, "Telomere Length and Replicative Aging in Human Vascular Tissues," *Proceedings of the National Academy of Science, USA*, vol. 92 (1995), 11190–11194.

40. A summary of recent research in this area can be found in C. W. Greider and E. H. Blackburn, "Telomeres, Telomerase and Cancer," *Scientific American*, February 1996, 92–97.

6. The Rate of Living

41. J. W. McArthur and W. H. T. Baillie, "Metabolic Activity and Duration of Life. II. Metabolic Rates and Their Relation to Longevity in *Daphnia magna*," *Journal of Experimental Zoology* 53 (1929): 243–286.

42. *Baltimore Sun*, November 24, 1940.

43. R. Pearl, A. C. Sutton, and W. T. Howard, Jr., "Experimental Treatment of Cancer with Tuberculin," *Lancet* 216 (1929): 1078–1080.

44. Edwin B. Wilson to Isaiah Bowman (letter), April 26, 1935.

45. E. A. Ross, *The Changing Chinese* (New York: Century Co., 1912).

46. J. R. Slonaker, "The Normal Activity of the Albino Rat from Birth to Natural Death, Its Rate of Growth and Duration of Life," *Journal of Animal Behavior* 2 (1912): 20–42.

47. I. D. Riley and D. Lehmann, "The Demography of Papua New Guinea: Migration, Fertility, and Mortality Patterns," in *Human Biology*, ed. R. D. Attenborough and M. P. Alpers.

48. From S. N. Austad and K. E. Fischer, "Mammalian Aging, Metabolism, and Ecology: Evidence from the Bats and Marsupials," *Journal of Gerontology* 46 (1991): B47–B53.

49. G. A. Sacher, "Life Table Modification and Life Prolongation," in *Handbook of the Biology of Aging*, 1st ed., ed. C. E. Finch and L. Hayflick (New York: Van Nostrand, 1977).

50. See, for instance, R. McCarter, E. J. Masoro, and B. P. Yu, "Does Food Restriction Retard Aging by Reducing the Metabolic Rate?" *American Journal of Physiology* 248 (1985): E488–E490.

51. E. Trinkaus and P. Shipman, *The Neanderthals* (New York: Knopf, 1993).

52. S. N. Austad and K. E. Fischer, "Primate Longevity: Its Place in the Mammalian Scheme," *American Journal of Primatology* 28 (1992): 251–261.

7. What Evolution Explains about Aging

53. A discussion of the role of ApoE in heart disease can be found in P. W. F. Wilson et al., "Apolipoprotein E Alleles, Dyslipidemia, and Coronary Heart Disease," *Journal of the American Medical Association* 272 (1994):

1666–1671. For its role in Alzheimer's disease, see A. M. Saunders et al., "Association of Apolipoprotein E Allele Epsilon 4 with Late-Onset Familial and Sporadic Alzheimer's Disease," *Neurology* 43 (1993): 1467–1472.

54. Grossman, "Interactions," 257–261.

55. J. B. Hamilton and G. B. Mestler, "Mortality and Survival: Comparison of Eunuches with Intact Men and Women in a Mentally Retarded Population," *Journal of Gerontology* 24 (1969): 395–411.

56. R. Schneider, "Comparison of Age, Sex, and Incidence Rates in Human and Canine Breast Cancer," *Cancer* 26 (1970): 419–426.

57. See R. T. Bronson, "Age at Death of Necropsied Intact and Neutered Cats," *American Journal of Veterinary Research* 42 (1981): 1606–1608, and "Variation in Age at Death of Dogs of Different Sexes and Breeds," *American Journal of Veterinary Research* 43 (1981): 2057–2059.

58. Research on the evolution and genetics of aging in fruit flies, as well as a detailed and highly technical explanation of the evolution of aging, can be found in M. R. Rose, *Evolutionary Biology of Aging* (Oxford: Oxford University Press, 1991).

59. Amazingly, at about the same time an independent team of investigators began an almost identical experiment, which got almost identical results and thus provided almost immediate confirmation of the results. See L. S. Luckinbill et al., "Selection for Delayed Senescence in *Drosophila melanogaster*," *Evolution* 38 (1984): 996–1003. And interestingly, these two independently created long-lived fruit flies differ in a number of aspects of their physiology, emphasizing that there are a multitude of aging processes and life may be extended significantly by altering any number of these processes.

60. The human history of Sapelo Island is told in W. S. McFeely, *Sapelo's People* (New York: Norton & Co., 1994).

61. A more staid scientific account of this research can be found in S. N. Austad, "Retarded Senescence in an Insular Population of Virginia Opossums (*Didelphis virginiana*)," *Journal of Zoology* 229 (1993): 695–708.

8. What Processes Cause Aging?

62. Eddie Andrade's story is masterfully told by Michael Fessier, Jr., "Death of a Fighter," *Los Angeles Times Magazine*, October 29, 1995.

63. An intelligible summary of the suspected role of mitochondrial DNA damage and aging can be found in "Mitochondrial DNA May Hold a Key to

Human Degenerative Diseases," *The Journal of NIH Research* 4 (June 1992): 62–66.

64. A number of Bruce Ames's papers are perfectly understandable by lay readers. Two of the most accessible are "Oxidants, Antioxidants, and the Degenerative Diseases of Aging," in the *Proceedings of the National Academy of Sciences, USA*, vol. 90 (September 1, 1993), 7915–7922, and "Understanding the Causes of Aging and Cancer," *Microbiologia* vol. 11 (September 11, 1995), 305–308. Dr. Ames is also an eloquent advocate for the consumption of lots of dietary antioxidants by eating fruit and vegetables.

65. A detailed description of this process that is not overly technical can be found in N. Woolf and M. J. Davies, "Arterial Plaque and Thrombus Formation," *Science and Medicine*, vol. 1 (September/October 1994), 38–47. A more technical account is given in R. Ross, "The Pathogenesis of Atherosclerosis: A Perspective for the 1990s," *Nature* 362 (1993): 801–809.

66. This rare disease, called chronic granulomatous disease, not only makes its sufferers susceptible to recurrent infections but also causes widespread pimplelike sores.

67. A few reports of positive correlations between the level of antioxidants and species longevity have been reported (e.g., J. M. Tolmasoff et al., "Superoxide Dismutase: Correlation with Life Span and Specific Metabolic Rate in Primate Species," *Proceedings of the National Academy of Sciences, USA*, vol. 77 (1980), 2777–2781), but these results have come from a single laboratory and are not generally accepted by other researchers in the field (for instance, see R. S. Sohal, "The Free Radical Hypothesis of Aging: An Appraisal of the Current Status," *Aging, Clinical and Experimental Research* 5 [1993]: 3–17).

68. The "worm" research is summarized in T. E. Johnson et al., "Genetics of Aging and Longevity in Lower Organisms," in *Cellular Aging and Cell Death*, ed. N. Holbrook and G. R. Martin (New York: John Wiley & Sons, 1996), and the genetically engineered fruit fly research is described in W. C. Orr and R. S. Sohal, "Extension of Life-Span by Overexpression of Superoxide Dismutase and Catalase in *Drosophila melanogaster*," *Science*, vol. 263 (1994), 1128–1130.

69. R. S. Sohal et al., "Oxidative Damage, Mitochondrial Oxidant Generation and Antioxidant Defenses during Aging and in Response to Food Restriction in the Mouse," *Mechanisms of Ageing and Development* 74 (1994): 121–133.

70. For a very readable account of the impact of glucose on aging, see A. Cerami et al., "Glucose and Aging," *Scientific American*, vol. 256 (May 1987), 90–96.

71. M. A. Smith et al., "Advanced Maillard Reaction End Products Are Associated with Alzheimer Disease Pathology," *Proceedings of the National Academy of Science, USA*, vol. 91 (1994), 5710–5714.

72. E. J. Masoro et al., "Evidence for the Glycation Hypothesis of Aging from the Food-Restricted Rodent Model," *Journal of Gerontology* 44 (1989): B20–B22.

73. Although there is no relationship among blood glucose, body temperature, and longevity, there is some evidence that long-lived mammals accumulate browning products more slowly than do short-lived animals. See D. R. Sell et al., "Longevity and the Genetic Determination of Collagen Glycoxidation Kinetics in Mammalian Senescence," *Proceedings of the National Academy of Sciences, USA*, vol. 93 (1996), 485–490.

74. The herb thyme seems particularly active in this sense. See Y. Morimitsu et al., "Protein Glycation Inhibitors from Thyme," *Bioscience, Biotechnology, and Biochemistry* 59 (1995): 2018–2021. Also, a synthetic drug called aminoguanidine is currently undergoing clinical trials for the treatment of diabetic kidney damage, and has shown promise in animal studies of being effective in limiting diabetes-related blindness.

75. For a technical discussion of this point, see B. S. Kristal and B. P. Yu, "An Emerging Hypothesis: Synergistic Induction of Aging by Free Radicals and Maillard Reactions," *Journal of Gerontology: Biological Sciences* 47 (1992): B107–B114.

76. The story of John Harrison's efforts to win a king's ransom by solving the "longitude problem" is grippingly told in Dava Sobel, *Longitude* (Walker & Co., 1995).

9. *Reproductive Aging, Menopause, and Health*

77. P. M. Waser, "Postreproductive Survival and Behavior in a Free-Ranging Female Mangabey," *Folia Primatologica* 29 (1978): 142–160.

78. K. Hill and A. M. Hurtado, "Hunter-Gatherers of the New World," *American Scientist*, vol. 77 (1989) 437–443, and idem, "The Evolution of Premature Reproductive Senescence and Menopause in Human Females: An Evaluation of the 'Grandmother Hypothesis,'" *Human Nature* 2 (1991): 313–350.

79. P. M. Sarrel, E. G. Lufkin, M. J. Oursler, and D. Keefe, "Estrogen Actions in Arteries, Bone, and Brain," *Scientific American Science and Medicine* (July/August 1994), 44–53.

80. There is a large body of literature on the topic of hormones and gynecological cancers. A couple of general papers on the topic are T. J. A. Key

and M. C. Pike, "The Role of Oestrogens and Progestagens in the Epidemiology and Prevention of Breast Cancer," *European Journal of Cancer and Clinical Oncology* 24: 29–43 and B. E. Henderson et al., "Endogenous Hormones as a Major Factor in Human Cancer," *Cancer Research* 42 (1982): 3232–3239.

81. P. T. Ellison, "Human Reproductive Ecology," *Annual Review of Anthropology*, vol. 23 (1994), 255–275.

82. A clear discussion of the pitfalls and practices of risk-factor epidemiology can be found in *Science*, vol. 269 (July 14, 1995), 164–169.

83. R. P. Feynman, *Six Easy Pieces* (Reading, Mass.: Addison-Wesley, 1995).

84. D. A. Savitz, H. Wachtel, F. A. Barnes, E. M. John, and J. G. Tvrdik, "Case-Control Study of Childhood Cancer and Exposure to 60-Hz Magnetic Fields," *American Journal of Epidemiology* 128 (1988): 21–38.

85. Editorial from the United Kingdom: "500 a Day Seek Help on Osteoporosis," *Medical News* 4 (1987): 8.

86. Estimates of the cost of hip fractures range as high as $18 billion per year and are expected to triple over the next 25 years. See L. V. Avioli, "Impact of the Menopause on Skeletal Metabolism and Osteoporotic Syndromes," *Experimental Gerontology* 29 (1994): 391–415.

87. Y. Beyene, *From Menarche to Menopause: Reproductive Lives of Peasant Women in Two Cultures* (Albany, N.Y.: State University of New York Press, 1986).

88. M. Lock, "Menopause in Cultural Context," *Experimental Gerontology* 29 (1994): 307–317.

89. B. Ettinger et al., "Reduced Mortality Associated with Long-Term Postmenopausal Estrogen Therapy," *Obstetrics & Gynecology* 87 (1996): 6–12.

90. T. Gura, "Estrogen: Key Player in Heart Disease Among Women," *Science*, vol. 269 (1995), 771–773.

91. This study is called the PEPI (Postmenopausal Estrogen/Progestin Interventions) trial. Its report, "Effects of Hormone Therapy on Bone Mineral Density," can be found in *Journal of the American Medical Association* 276 (November 1996): 1389–1396. In the same issue, there is a report by a second, but shorter-term, clinical trial, and an editorial on estrogen's postmenopausal effects.

92. D. T. Felson et al., "The Effect of Postmenopausal Estrogen Therapy on Bone Density in Elderly Women," *New England Journal of Medicine* 329 (1993): 1141–1146.

93. H. K. Genant, D. J. Baylink, and J. C. Gallagher, "Estrogens in the Prevention of Osteoporosis in Postmenopausal Women," *American Journal of Obstetrics & Gynecology* 161 (1989): 1842–1846.

94. M. Breckwoldt, C. Keck, and U. Karck, "Benefits and Risks of Hormone Replacement Therapy," *Journal of Steroid Biochemistry & Molecular Biology* 53 (1995): 205–208.

95. P. A. Newcomb and B. A. Storer, "Postmenopausal Hormone Use and Risk of Large-Bowel Cancer," *Journal of the National Cancer Institute* 87 (1995): 1067–1071.

96. Current reports differ widely on whether this effect exists. For diametrically opposite conclusions, see V. M. Henderson et al., "Estrogen Replacement Therapy in Older Women: Comparisons between Alzheimer's Disease Cases and Nondemented Control Subjects," *Archives of Neurobiology* 51 (1994): 896–900, and D. E. Brenner et al., "Postmenopausal Estrogen Replacement Therapy and the Risk of Alzheimer's Disease: A Population-based Case-Control Study," *American Journal of Epidemiology* 140 (1994): 262–267.

97. There is an excellent series of articles discussing the risks and benefits of hormone-replacement therapy, including the different effects of taking hormones via pills versus by transdermal skin patches in the September 1995 issue of the *American Journal of Obstetrics & Gynecology*.

98. P. A. Newcomb et al., "Long-term Hormone Replacement Therapy and Risk of Breast Cancer in Postmenopausal Women," *American Journal of Epidemiology* 142 (1995): 788–795.

10. Slowing Aging and Extending Life: Remedies and Expectations

99. More information on the amusing history of bogus antiaging therapies can be found in Roger Gosden's *Cheating Time: Science, Sex, and Ageing* (London: Macmillan, 1996) and Leonard Hayflick's *How and Why We Age* (New York: Ballantine, 1994). A general assessment of research on Gerovital H3 can be found in A. Ostfeld et al., "The Systemic Use of Procaine in the Treatment of the Elderly: A Review," *Journal of the American Geriatrics Society* 25 (1977): 1–19.

100. See notes 69 and 72 for the appropriate references to reduced free-radical production and blood glucose.

101. For a highly readable discussion of more current research on this issue, see Robert Sapolsky's *Why Zebras Don't Get Ulcers* (New York: Freeman & Company, 1994).

102. For various sides of this debate from a technical perspective, see J. F. Nelson et al., "Neuroendocrine Involvement in Aging: Evidence from Studies of Reproductive Aging and Caloric Restriction," *Neurobiology of Aging* 16 (1995): 837–843, and the various commentaries immediately following that article.

103. Hormesis and these early experiments are discussed in George Sacher, "Life Table Modification and Life Prolongation," in *Handbook of the Biology of Aging*, ed. C. E. Finch and L. Hayflick, 1st ed. (New York: Van Nostrand, 1977).

104. Compare, for instance, S. F. Loy et al., "Effects of 24-Hour Fast on Cycling Endurance Time at Two Different Intensities," *Journal of Applied Physiology* 61 (1986): 654–659, and G. L. Dohm et al., "Influence of Fasting on Glycogen Depletion in Rats during Exercise," *Journal of Applied Physiology* 55 (1983): 830–833.

105. See *Dietary Restriction: Implications for the Design and Interpretation of Toxicity and Carcinogenicity Studies*, ed. R. W. Hart, D. A. Neumann, and R. T. Robertson (Washington, D.C.: ILSA Press, 1995).

106. M. M. Lee et al., "Comparison of Dietary Habits, Physical Activity, and Body Size among Chinese in North America and China," *International Journal of Epidemiology* 23 (1994): 984–990.

107. To calculate your BMI from American measures, use the following formula:

$$\frac{704.5 \times \text{Weight (in pounds)}}{\text{Height}^2 \text{ (in inches)}}$$

108. Department of Agriculture, Department of Health and Human Services, *Nutrition and Your Health: Dietary Guidelines for Americans*, 3rd ed. (Washington, D.C.: U.S. Government Printing Office).

109. For different points of view on this matter, see T. Harris, E. F. Cook et al., "Body Mass Index and Mortality among Nonsmoking Older Persons: The Framingham Heart Study," *Journal of the American Medical Association* 259 (1988): 1520–1524, versus M. Schroll, "A Longitudinal Epidemiological Survey of Relative Weight at age 25, 50, and 60 in the Glostrup Population of Men and Women Born in 1914," *Danish Medical Bulletin*, vol. 28 (1981), 106–116, versus T. Wilcosky et al., "Obesity and Mortality in the Lipid Research Clinics Program Follow-up Study," *Journal of Clinical Epidemiology* 43 (1990): 743–752.

110. J. E. Manson et al., "Body Weight and Mortality among Women," *New England Journal of Medicine* 333 (1995): 677–687.

111. Comparing studies from the National Institute of Health's Gerontology Research Center and the Wisconsin Regional Primate Research Center, control animals of roughly equivalent ages weighed 9 and 14 kilograms, respectively, compared with food-restricted animals, which weighed 7 and 9.5 kilograms, respectively. See R. Weindruch et al., "Measures of Body Size and Growth in Rhesus and Squirrel Monkeys Subjected to Long-term Dietary Restriction," *American Journal of Primatology* 35 (1995): 207–228, and J. W. Kemnitz et al., "Dietary Restriction Increases Insulin Sensitivity and Lowers Blood Glucose in Rhesus Monkeys," *American Journal of Physiology* 266 (1994): E540–E547.

112. I thank Akira Homma of the Tokyo Metropolitan Institute of Gerontology for providing me with Japanese longevity data as of 1994.

113. There is a hint that the immune systems of long-lived Okinawans may be genetically superior in some way. See H. Takata et al., "Influence of Major Histocompatibility Complex Region Genes on Human Longevity among Okinawan-Japanese Centenarians and Nonagenarians," *The Lancet* (October 10, 1987), 824–826.

114. R. L. Walford, S. B. Harris, and M. W. Gunion, "The Calorically Restricted Low-Fat, Nutrient-Dense Diet in Biosphere 2 Significantly Lowers Blood Glucose, Total Leukocyte Count, Cholesterol, and Blood Pressure in Humans," *Proceedings of the National Academy of Science, USA,* vol. 89 (1992), 11533–11537.

115. R. J. Kuczmarski, K. M. Flegal, S. M. Campbell, and C. L. Johnson, "Increasing Prevalence of Overweight among U.S. Adults: The National Health and Nutrition Examination Surveys. 1960–1991," *Journal of the American Medical Association* 272 (1991): 205–211.

116. A. Alling and M. Nelson, *Life under Glass* (Oracle, Ariz.: The Biosphere Press, 1993). Also, for a general account of the ecological problems that developed inside the Biosphere, see J. E. Cohen and D. Tilman, "Biosphere 2 and Biodiversity: The Lessons So Far," *Science,* vol. 274 (1996), 1150–1151.

117. A superb, though somewhat technical, discussion of this point can be found in J. O. Holloszy and W. M. Kohrt, "Exercise," in *Handbook of Physiology, Section 11: Aging,* ed. Edward J. Masoro (New York: Oxford University Press, 1995).

118. Scientific analyses of exercise and longevity from the Harvard Alumni Health study can be found in: (1) R. S. Paffenbarger et al., "Physical Activity, All-Cause Mortality, and Longevity of College Alumni," *New England Journal of Medicine* 314 (1986): 605–613; (2) idem, "The Association of Changes in Physical-Activity Level and Other Lifestyle Char-

acteristics with Mortality among Men," *New England Journal of Medicine* 328 (1993): 538–545; and (3) I.-M. Lee et al., "Exercise Intensity and Longevity in Men," *Journal of the American Medical Association* 273 (1995): 1179–1184.

119. A good summary of these animal studies is in B. P. Yu, "Putative Interventions" in *Handbook of Physiology, Section 11: Aging*, ed. Edward J. Masoro (New York: Oxford University Press, 1995).

120. A nice hatchet job was done on vitamin C and colds by T. Chalmers in 1975. This doesn't mean that Chalmers was wrong. The prevailing current opinion is that vitamin C does not prevent colds but may have a small effect in reducing the length and severity of symptoms. For a critique of Chalmers's paper, see H. Hemila and Z. S. Herman, "Vitamin C and the Common Cold: A Retrospective Analysis of Chalmers' Review," *Journal of the American College of Nutrition* 14 (1995): 116–123. However, note that one author of this paper is employed by the Linus Pauling Institute of Science and Medicine.

121. Specifically, vitamin C interferes with glucose, uric acid, creatinine, and inorganic phosphate values in standard blood-chemistry tests. People taking large supplements—say 1,000 milligrams or more per day—should inform their doctor before getting diagnostic blood tests.

122. One commercial product—*Spirulina*, a cyanobacterium—is often recommended by suppliers as a protein source and vitamin supplement. One spot check of nine samples of *Spirulina* found high concentrations of lead, copper, iron, manganese, and zinc, and a "higher than prudent" level of mercury, according to FDA and WHO guidelines. See P. E. Johnson and L. E. Shubert, "Accumulation of Mercury and Other Elements by Spirulina (Cyanophyceae)," *Nutrition Reports International*, vol. 34 (1986), 1063–1070. Impurities in another superstar food supplement of a few years ago—tryptophan—even led to a number of deaths.

123. Most high-dose, correctly performed (i.e., double blind with placebo controls) studies of vitamin C have been done on sick people to determine whether vitamin C could alleviate their illness (it generally didn't). The fact that no discernible side effects were observed in patients with advanced cancer taking 10,000 milligrams per day for a year does not necessarily mean longer use in healthy people will not lead to significant side effects. Also, some researchers make claims about the effects of vitamin C from studies that are really about diets rich in fruit and vegetables, assuming that any effect seen must be due to vitamin C—a fallacious assumption. The tendency of interested parties to selectively cite data favorable to their interpretation, and ignore similar studies finding the opposite effects is particularly rife in the vitamin C business.

124. V. Herbert et al., "Vitamin C-driven Free Radical Generation from Iron," *Journal of Nutrition* 126 (1996): 1213S–1220S.

125. Recent and fair reviews of many studies may be found in T. Byers and N. Guerrero, "Epidemiologic Evidence for Vitamin C and Vitamin E in Cancer Prevention," *American Journal of Nutrition* 62 (1995): 1385S–1392S, and E. W. Flagg et al., "Epidemiologic Studies of Antioxidants and Cancer in Humans," *Journal of the American College of Nutrition* 15 (1995): 419–427.

126. The three trials discussed are available as: The Alpha-tocopherol, Beta-carotene Cancer Prevention Study Group, "The Effect of Vitamin E and Beta Carotene on the Incidence of Lung Cancer and Other Cancers in Male Smokers," *New England Journal of Medicine* 330 (1994): 1029–1035; G. S. Omenn et al., "Effects of a Combination of Beta Carotene and Vitamin A on Lung Cancer and Cardiovascular Disease," *New England Journal of Medicine* 334 (1996): 1150–1155; and C. H. Hennekens et al., "Lack of Effect of Long-term Supplementation with Beta Carotene on the Incidence of Malignant Neoplasms and Cardiovascular Disease," *New England Journal of Medicine* 334 (1996): 1145–1149. A more recent follow-up study of the people in the first study also found no effect of beta-carotene on the incidence of angina pectoris, a sign of insufficient oxygen supply to the heart: J. M. Rapola et al., "Effect of Vitamin E and Beta Carotene on the Incidence of Angina Pectoris," *Journal of the American Medical Association* 275 (1996): 693–698.

127. W. Pierpaoli and W. Regelson, *The Melatonin Miracle* (New York: Simon & Schuster; London: Fourth Estate, 1995).

128. Fred Turek puts the various claims for melatonin into reasonable perspective in "Melatonin Hype Hard to Swallow," *Nature*, vol. 379 (1996), 295–296.

129. R. J. Reiter and J. Robinson, *Melatonin* (New York: Bantam, 1995).

130. These responsible contraindications and their rationale are listed in ibid., 205–206.

131. A readable, reasonably well-balanced treatment of DHEA as discussed at a recent conference hosted by the New York Academy of Sciences is B. Bilger, "Forever Young," *The Sciences* (September–October 1995), 26–30. Also, see readers' letters pertaining to this article in the May–June 1996 issue of the same journal.

132. A technical summary of deprenyl studies can be found in Holloszy and Kohrt, "Exercise," in *Handbook of Physiology*.

133. *After Many a Summer* (1939).

134. *Brave New World* (1932).

135. Some of the latest technical findings on this gene product, leptin, can be found in M. Maffei et al., "Leptin Levels in Human and Rodent: Measurement of Plasma Leptin and ob RNA in Obese and Weight-reduced Subjects," *Nature Genetics* 1 (1995): 1155–1161.

136. The new products, and a general discussion of the latest genetic obesity research, can be found in W. T. Gibbs, "Gaining on Fat," *Scientific American*, August 1996, 88–94.

137. See Orr and Sohal, "Extension of Life-Span," *Science*, 1128–1130.

Index